PE
civil engineering

geotechnical

practice exam

Copyright © 2014 by NCEES®. All rights reserved.

All NCEES sample questions and solutions are copyrighted under the laws of the United States. No part of this publication may be reproduced, stored in a retrieval system, or transmitted in any form or by any means without the prior written permission of NCEES. Requests for permissions should be addressed in writing to permissions@ncees.org or to NCEES Exam Publications, PO Box 1686, Clemson, SC 29633.

ISBN 978-1-932613-71-1

Printed in the United States of America
October 2014 First Printing

CONTENTS

Introduction to NCEES Exams..1
About NCEES
Updates on exam content and procedures
Exam-day schedule
Admission to the exam site
Examinee Guide
Scoring and reporting
Staying connected

Exam Specifications..3

Civil AM Practice Exam..11
Geotechnical PM Practice Exam..33

Civil AM Solutions..63
Geotechnical PM Solutions..83

About NCEES
The National Council of Examiners for Engineering and Surveying (NCEES) is a nonprofit organization made up of engineering and surveying licensing boards from all U.S. states and territories. Since its founding in 1920, NCEES has been committed to advancing licensure for engineers and surveyors in order to protect the health, safety, and welfare of the American public.

NCEES helps its member licensing boards carry out their duties to regulate the professions of engineering and surveying. It develops best-practice models for state licensure laws and regulations and promotes uniformity among the states. It develops and administers the exams used for engineering and surveying licensure throughout the country. It also provides services to help licensed engineers and surveyors practice their professions in other U.S. states and territories.

Updates on exam content and procedures
Visit us at **ncees.org/exams** for updates on everything exam-related, including specifications, exam-day policies, scoring, and corrections to published exam preparation materials. This is also where you will register for the exam and find additional steps you should follow in your state to be approved for the exam.

Exam-day schedule
Be sure to arrive at the exam site on time. Late-arriving examinees will not be allowed into the exam room once the proctor has begun to read the exam script. The report time for the exam will be printed on your Exam Authorization. Normally, you will be given 1 hour between morning and afternoon sessions.

Admission to the exam site
To be admitted to the exam, you must bring two items: (1) your Exam Authorization and (2) a current, signed, government-issued identification.

Examinee Guide
The *NCEES Examinee Guide* is the official guide to policies and procedures for all NCEES exams. All examinees are required to read this document before starting the exam registration process. You can download it at ncees.org/exams. It is your responsibility to make sure that you have the current version.

NCEES exams are administered in either a computer-based format or a pencil-and-paper format. Each method of administration has specific rules. This guide describes the rules for each exam format. Refer to the appropriate section for your exam.

Scoring and reporting
NCEES typically releases exam results to its member licensing boards 8–10 weeks after the exam. Depending on your state, you will be notified of your exam result online through your MyNCEES account or via postal mail from your state licensing board. Detailed information on the scoring process can be found at ncees.org/exams.

Staying connected
To keep up to date with NCEES announcements, events, and activities, connect with us on your preferred social media network.

EXAM SPECIFICATIONS

NCEES Principles and Practice of Engineering
CIVIL BREADTH and GEOTECHNICAL DEPTH
Exam Specifications

Effective Beginning with the April 2015 Examinations

- The civil exam is a breadth and depth exam. This means that examinees work the breadth (AM) exam and one of the five depth (PM) exams.

- The five areas covered in the civil exam are construction, geotechnical, structural, transportation, and water resources and environmental. The breadth exam contains questions from all five areas of civil engineering. The depth exams focus more closely on a single area of practice in civil engineering.

- Examinees work all questions in the morning session and all questions in the afternoon module they have chosen. Depth results are combined with breadth results for final score.

- The exam is an 8-hour open-book exam. It contains 40 multiple-choice questions in the 4-hour AM session, and 40 multiple-choice questions in the 4-hour PM session.

- The exam uses both the International System of Units (SI) and the US Customary System (USCS).

- The exam is developed with questions that will require a variety of approaches and methodologies, including design, analysis, and application.

- The examples specified in knowledge areas are not exclusive or exhaustive categories.

- The specifications for the **AM exam** and the **Geotechnical PM exam** are included here. The **design standards** applicable to the Geotechnical PM exam are shown on **ncees.org**.

CIVIL BREADTH Exam Specifications

		Approximate Number of Questions
I.	**Project Planning**	4
	A. Quantity take-off methods	
	B. Cost estimating	
	C. Project schedules	
	D. Activity identification and sequencing	
II.	**Means and Methods**	3
	A. Construction loads	
	B. Construction methods	
	C. Temporary structures and facilities	
III.	**Soil Mechanics**	6
	A. Lateral earth pressure	
	B. Soil consolidation	
	C. Effective and total stresses	
	D. Bearing capacity	
	E. Foundation settlement	
	F. Slope stability	

Civil–Breadth Exam Specifications Continued

IV. Structural Mechanics — 6
  A. Dead and live loads
  B. Trusses
  C. Bending (e.g., moments and stresses)
  D. Shear (e.g., forces and stresses)
  E. Axial (e.g., forces and stresses)
  F. Combined stresses
  G. Deflection
  H. Beams
  I. Columns
  J. Slabs
  K. Footings
  L. Retaining walls

V. Hydraulics and Hydrology — 7
  A. Open-channel flow
  B. Stormwater collection and drainage (e.g., culvert, stormwater inlets, gutter flow, street flow, storm sewer pipes)
  C. Storm characteristics (e.g., storm frequency, rainfall measurement and distribution)
  D. Runoff analysis (e.g., Rational and SCS/NRCS methods, hydrographic application, runoff time of concentration)
  E. Detention/retention ponds
  F. Pressure conduit (e.g., single pipe, force mains, Hazen-Williams, Darcy-Weisbach, major and minor losses)
  G. Energy and/or continuity equation (e.g., Bernoulli)

VI. Geometrics — 3
  A. Basic circular curve elements (e.g., middle ordinate, length, chord, radius)
  B. Basic vertical curve elements
  C. Traffic volume (e.g., vehicle mix, flow, and speed)

VII. Materials — 6
  A. Soil classification and boring log interpretation
  B. Soil properties (e.g., strength, permeability, compressibility, phase relationships)
  C. Concrete (e.g., nonreinforced, reinforced)
  D. Structural steel
  E. Material test methods and specification conformance
  F. Compaction

Civil–Breadth Exam Specifications Continued

VIII. Site Development   5
    A. Excavation and embankment (e.g., cut and fill)
    B. Construction site layout and control
    C. Temporary and permanent soil erosion and sediment control (e.g., construction erosion control and permits, sediment transport, channel/outlet protection)
    D. Impact of construction on adjacent facilities
    E. Safety (e.g., construction, roadside, work zone)

CIVIL–GEOTECHNICAL Depth Exam Specifications

Approximate Number of Questions

I. **Site Characterization** — 5
 A. Interpretation of available existing site data and proposed site development data (e.g., aerial photography, geologic and topographic maps, GIS data, as-built plans, planning studies and reports)
 B. Subsurface exploration planning
 C. Geophysics (e.g., GPR, resistivity, seismic methods)
 D. Drilling techniques (e.g., hollow stem auger, cased boring, mud rotary, air rotary, rock coring, sonic drilling)
 E. Sampling techniques (e.g., split-barrel sampling, thin-walled tube sampling, handling and storage)
 F. In situ testing (e.g., standard penetration testing, cone penetration testing, pressure meter testing, dilatometer testing, field vane shear)
 G. Description and classification of soils (e.g., Burmeister, Unified Soil Classification System, AASHTO, USDA)
 H. Rock classification and characterization (e.g., recovery, rock quality designation, RMR, weathering, orientation)
 I. Groundwater exploration, sampling, and characterization

II. **Soil Mechanics, Laboratory Testing, and Analysis** — 5
 A. Index properties and testing
 B. Strength testing of soil and rock
 C. Stress-strain testing of soil and rock
 D. Permeability testing properties of soil and rock
 E. Effective and total stresses

III. **Field Materials Testing, Methods, and Safety** — 3
 A. Excavation and embankment, borrow source studies, laboratory and field compaction
 B. Trench and construction safety
 C. Geotechnical instrumentation (e.g., inclinometer, settlement plates, piezometer, vibration monitoring)

IV. **Earthquake Engineering and Dynamic Loads** — 2
 A. Liquefaction analysis and mitigation techniques
 B. Seismic site characterization, including site classification using ASCE 7
 C. Pseudo-static analysis and earthquake loads

V. **Earth Structures** — 4
 A. Slab on grade
 B. Ground improvement (e.g., grouting, soil mixing, preconsolidation/wicks, lightweight materials)
 C. Geosynthetic applications (e.g., separation, strength, filtration, drainage, reinforced soil slopes, internal stability of MSE)
 D. Slope stability and slope stabilization
 E. Earth dams, levees, and embankments
 F. Landfills and caps (e.g., interface stability, drainage systems, lining systems)

Civil–Geotechnical Depth Exam Specifications Continued

- G. Pavement structures (rigid, flexible, or unpaved), including equivalent single-axle load (ESAL), pavement thickness, subgrade testing, subgrade preparation, maintenance and rehabilitation treatments
- H. Settlement

VI. Groundwater and Seepage   3
- A. Seepage analysis/groundwater flow
- B. Dewatering design, methods, and impact on nearby structures
- C. Drainage design/infiltration
- D. Grouting and other methods of reducing seepage

VII. Problematic Soil and Rock Conditions   3
- A. Karst; collapsible, expansive, and sensitive soils
- B. Reactive/corrosive soils
- C. Frost susceptibility

VIII. Earth Retaining Structures (ASD or LRFD)   5
- A. Lateral earth pressure
- B. Load distribution
- C. Rigid retaining wall stability analysis (e.g., CIP, gravity, external stability of MSE, crib, bin)
- D. Flexible retaining wall stability analysis (e.g., soldier pile and lagging, sheet pile, secant pile, tangent pile, diaphragm walls, temporary support of excavation, braced and anchored walls)
- E. Cofferdams
- F. Underpinning (e.g., effects on adjacent construction)
- G. Ground anchors, tie-backs, soil nails, and rock anchors for foundations and slopes

IX. Shallow Foundations (ASD or LRFD)   5
- A. Bearing capacity
- B. Settlement, including vertical stress distribution

X. Deep Foundations (ASD or LRFD)   5
- A. Single-element axial capacity (e.g., driven pile, drilled shaft, micropile, helical screw piles, auger cast piles)
- B. Lateral load and deformation analysis
- C. Single-element settlement
- D. Downdrag
- E. Group effects (e.g., axial capacity, settlement, lateral deflection)
- F. Installation methods/hammer selection
- G. Pile dynamics (e.g., wave equation, high-strain dynamic testing, signal matching)
- H. Pile and drilled-shaft load testing
- I. Integrity testing methods (e.g., low-strain impact integrity testing, ultrasonic cross-hole testing, coring, thermal integrity testing)

CIVIL AM PRACTICE EXAM

CIVIL AM PRACTICE EXAM

101. A 227-ft length of canal is to be lined with concrete for erosion control. With 12% allowance for waste and overexcavation, the volume (yd^3) of concrete that must be delivered is most nearly:

(A) 234
(B) 280
(C) 292
(D) 327

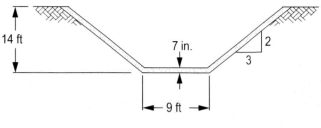

LINED LENGTH = 227 ft

102. Based on the straight-line method of depreciation, the book value at the end of the 8th year for a track loader having an initial cost of $75,000, and a salvage value of $10,000 at the end of its expected life of 10 years is most nearly:

(A) $10,000
(B) $15,000
(C) $23,000
(D) $48,750

103. The budgeted labor amount for an excavation task is $4,000. The hourly labor cost is $50 per worker, and the workday is 8 hours. Two workers are assigned to excavate the material. The time (days) available for the workers to complete this task is most nearly:

(A) 3
(B) 4
(C) 5
(D) 12.5

CIVIL AM PRACTICE EXAM

104. A CPM arrow diagram is shown below. Nine activities have been estimated with durations ranging from 5 to 35 days. The minimum time (days) required to finish the project is most nearly:

(A) 40
(B) 42
(C) 45
(D) 50

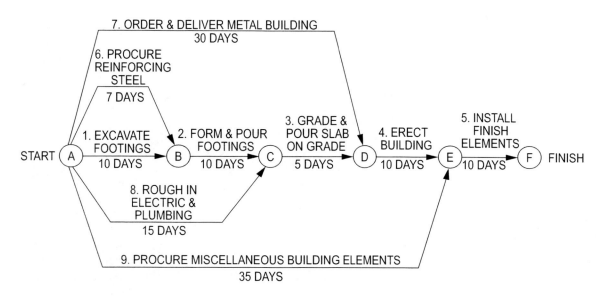

105. A bridge is to be jacked up to replace its bearings. The design requires a hydraulic ram with a minimum capacity of 1,000 kN (kilonewtons). The hydraulic rams that are available are rated in tons (2,000 lb/ton). The **minimum** size (tons) ram to use is most nearly:

(A) 1,110
(B) 250
(C) 150
(D) 100

106. A crane with a 100-ft boom is being used to set a small load on the roof of the building shown. The minimum standoff (Point A) from the corner of the building to the centerline of the boom is indicated. What is the maximum distance (ft) from the edge of the building that the load can be placed on the roof?

(A) 16
(B) 25
(C) 30
(D) 36

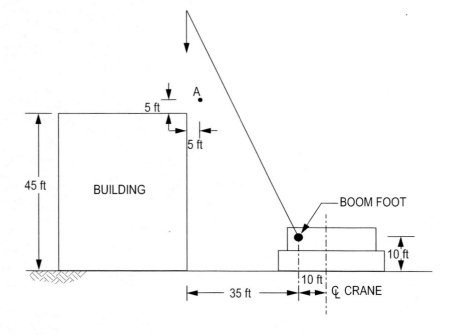

107. A wall form subjected to a wind load of 20 psf is prevented from overturning by diagonal braces spaced at 8 ft on center along the length of the wall form as shown in the figure. The connection at the base of the form at Point A is equivalent to a hinge. Ignore the weight of the form. The axial force (lb) resisted by the brace is most nearly:

(A) 2,050
(B) 2,560
(C) 2,900
(D) 4,525

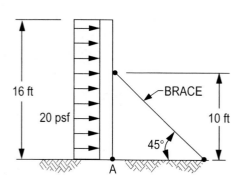

CIVIL AM PRACTICE EXAM

108. Which one of the following statements regarding lateral earth pressures is correct?

(A) The lateral strain required to fully mobilize the soil passive pressure is considerably smaller than the lateral strain required to fully mobilize the soil active pressure.

(B) The lateral strain required to fully mobilize the soil passive pressure is slightly smaller than the lateral strain required to fully mobilize the soil active pressure.

(C) The lateral strain required to fully mobilize the soil passive pressure is slightly greater than the lateral strain required to fully mobilize the soil active pressure.

(D) The lateral strain required to fully mobilize the soil passive pressure is considerably greater than the lateral strain required to fully mobilize the soil active pressure.

109. Site preparation and grading require the placement of 20 ft of new fill. An analysis of the resulting consolidation of the underlying soft, saturated, compressible deposits reveals a mean consolidation settlement of 22 in. affecting a 21.5-acre area. Prefabricated wick drains will be used to accelerate the settlement to meet the project schedule. Because of contamination from the former site use, the effluent from the wick drains will need to be collected and treated prior to disposal at an estimated cost of $0.25 per gallon. Assuming no loss of effluent during collection, the estimated treatment and disposal cost for the wick drain effluent at this site is most nearly:

(A) $430,000
(B) $3,200,000
(C) $5,200,000
(D) $35,000,000

110. A soil profile is shown in the figure. The effective vertical stress (psf) at Point A is most nearly:

(A) 1,270
(B) 1,820
(C) 2,140
(D) 2,570

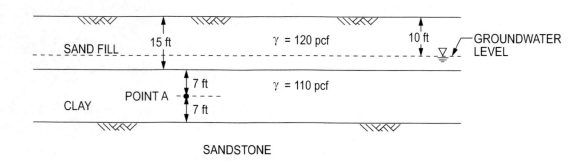

111. A bridge footing is to be constructed in sand. The groundwater level is at the ground surface. The ultimate bearing capacity would be based on what type of soil unit weight?

(A) Buoyant unit weight

(B) Saturated unit weight

(C) Dry unit weight

(D) Total unit weight

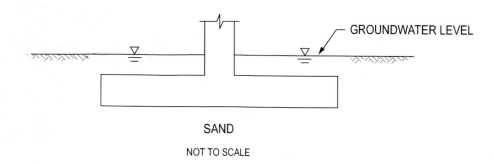

112. The figure shows two identical building footings with the same load but constructed in two different soil types. Which of the following statements is most correct?

(A) The long-term settlement for Case I is less than Case II.
(B) The long-term settlement for Case II is less than Case I.
(C) The long-term settlements are the same for both cases.
(D) Settlement is not a concern for either case.

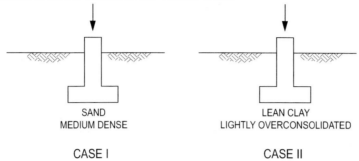

113. The minimum factor of safety against rotational failure for permanent slopes under long-term, non-seismic conditions influencing occupied structures is closest to:

(A) 1.0
(B) 1.1
(C) 1.5
(D) 3.0

114. Referring to the figure, what load combination produces the maximum uplift on Footing A?

(A) Dead + live
(B) Dead + wind
(C) Dead
(D) Dead + live + wind

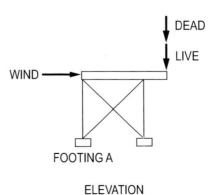

115. A simply supported truss is loaded as shown in the figure. The loads (kips) for Members b and c are most nearly:

	Member b	Member c
(A)	0	0
(B)	0	100
(C)	100	0
(D)	100	100

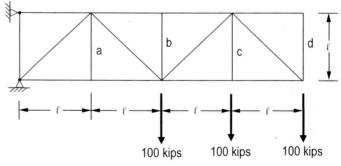

116. Consider two beams with equal cross-sections, made of the same material, having the same support conditions, and each loaded with equal uniform load per length. One beam is twice as long as the other. The maximum bending stress in the longer beam is larger by a factor of:

(A) 1.25
(B) 2
(C) 3
(D) 4

117. The point load (kips) placed at the centerline of a 30-ft beam that produces the same maximum shear in the beam as a uniform load of 1 kip/ft is most nearly:

(A) 7.5
(B) 15
(C) 30
(D) 60

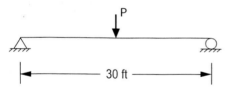

118. The beam sections shown are fabricated from 1/2-in. × 6-in. steel plates. Which of the following cross sections will provide the greatest flexural rigidity about the x-axis?

(A)

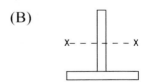

(B)

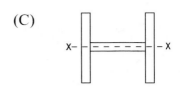

(C)

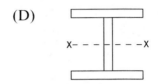

(D)

CIVIL AM PRACTICE EXAM

119. A concrete gravity retaining wall having a unit weight of 150 pcf is shown in the figure. Use the Rankine active earth pressure theory and neglect wall friction. The factor of safety against overturning about the toe at Point O is most nearly:

(A) 3.1
(B) 2.5
(C) 2.2
(D) 0.3

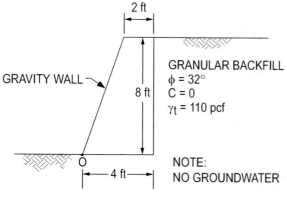

120. A drainage basin produces a stormwater runoff volume of 25.0 acre-ft, which must be drained through a rectangular channel that is 4 ft wide and 2 ft deep and has a uniform slope of 0.2%. Assume a Manning roughness coefficient of 0.022 and a constant depth of flow of 1.5 ft. The time (hours) it will take to discharge the runoff is most nearly:

(A) 12.5
(B) 16.4
(C) 18.5
(D) 25.0

CIVIL AM PRACTICE EXAM

121. Two identical 12-in. storm sewers flow full at a 2% slope into a junction box. A single larger pipe of the same material and slope flows out of the box. Assuming the following pipe sizes are commercially available, the minimum size of this downstream pipe (in.) designed to flow full is most nearly:

(A) 16
(B) 18
(C) 20
(D) 24

122. The following table represents the rainfall recorded from all rain gages located in and around a drainage area.

Gage	A	B	C	D	E	F	G	H	I	J	K
Rainfall (in.)	2.1	3.6	1.3	1.5	2.6	6.1	5.1	4.8	4.1	2.8	3.0

Using the arithmetic mean method, the average precipitation (in.) for the drainage area is most nearly:

(A) 3.4
(B) 3.7
(C) 4.1
(D) 37.0

123. The rational method must be used to determine the maximum runoff rate for a 90-acre downtown area. The time of concentration for the 50-year frequency storm is 1 hour. Intensity-duration-frequency curves and a table of runoff coefficients are provided. The maximum runoff rate (cfs), based on the maximum runoff coefficient for a 50-year storm, is most nearly:

(A) 160
(B) 220
(C) 300
(D) 340

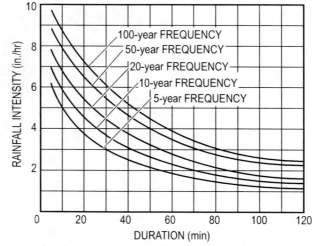

INTENSITY-DURATION-FREQUENCY CURVES

123. (Continued)

Description of Area	Runoff Coefficients
Business	
Downtown areas	0.70–0.95
Neighborhood areas	0.50–0.70
Residential	
Single-family areas	0.30–0.50
Multiunits, detached	0.40–0.60
Multiunits, attached	0.60–0.75
Residential (suburban)	0.25–0.40
Apartment dwelling areas	0.50–0.70
Industrial	
Light areas	0.50–0.80
Heavy areas	0.60–0.90
Parks, cemeteries	0.10–0.25
Playgrounds	0.20–0.35
Railroad yard areas	0.20–0.40
Unimproved areas	0.10–0.30
Streets	
Asphalt	0.70–0.95
Concrete	0.80–0.95
Brick	0.70–0.85
Drives and walks	0.75–0.85

CIVIL AM PRACTICE EXAM

124. A stormwater drainage ditch with a maximum capacity of 10 cfs discharges into a detention basin. The detention basin volume is 400,000 gal. During a storm event the average discharge into the detention basin was 1.5 cfs. The time (hours) to fill the empty basin would be most nearly:

(A) 1.5
(B) 9.9
(C) 11.1
(D) 74.1

125. Assume fully turbulent flow in a 1,650-ft section of 3-ft-diameter pipe. The Darcy-Weisbach friction factor f is 0.0115. There is a 5-ft drop in the energy grade line over the section. The flow rate (cfs) is most nearly:

(A) 16
(B) 29
(C) 50
(D) 810

126. Assuming that Bernoulli's equation applies (ignore head losses) to the pipe flow shown in the figure, which of the following statements is most correct?

(A) Pressure head increases from 1 to 2.
(B) Pressure head decreases from 1 to 2.
(C) Pressure head remains unchanged from 1 to 2.
(D) Bernoulli's equation does not include pressure head.

DIRECTION OF FLOW

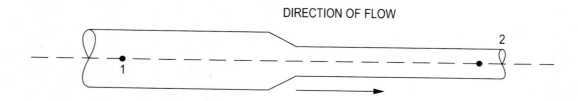

CIVIL AM PRACTICE EXAM

127. The following information is for a proposed horizontal curve in a new subdivision:

PI station   12+40.00
Degree of curve   10°
Deflection angle   12°30′

The station of the PT is most nearly:

(A) 12+79.80
(B) 12+80.10
(C) 13+02.00
(D) 13+64.75

128. For the sag vertical curve shown, the tangent slope at Station 14+00 is most nearly:

(A) +0.53%
(B) +1.23%
(C) +2.12%
(D) +2.77%

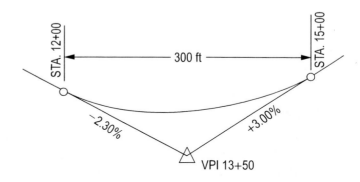

NOT TO SCALE

129. An interstate highway has the following traffic count data for a day in each month as shown below:

Jan.	63,500
Feb.	62,100
Mar.	64,400
Apr.	64,900
May	75,800
June	77,300
July	78,950
Aug.	77,200
Sept.	70,050
Oct.	69,000
Nov.	66,000
Dec.	64,000
Annual Total	833,200

The seasonal factor for the summer months of June through August is most nearly:

(A) 0.28
(B) 0.89
(C) 1.02
(D) 1.12

130. The most essential criteria for proper soil classification using the Unified Soil Classification or the AASHTO Soil Classification system are:

(A) water content and soil density
(B) Atterberg limits and specific gravity
(C) grain-size distribution and water content
(D) grain-size distribution and Atterberg limits

CIVIL AM PRACTICE EXAM

131. The Standard Penetration Test (SPT) is widely used as a simple and economic means of obtaining which of the following?

(A) A measurement of soil compressibility expressed in terms of a compression index

(B) A direct measurement of the undrained shear strength

(C) An indirect indication of the relative density of cohesionless soils

(D) A direct measurement of the angle of internal friction

132. A department of transportation must remove and replace a 12-ft × 20-ft concrete slab on an interstate facility. To minimize disruption to traffic, the work must be completed during an 8-hour nighttime work shift. Nighttime temperatures average 50°F. If the minimum required compressive strength is 3,500 psi, the concrete mix would most likely consist of:

(A) coarse aggregate, sand, Type II cement, chemical accelerator

(B) sand, Type III cement, water, chemical accelerator

(C) coarse aggregate, sand, Type V cement, water, chemical accelerator

(D) coarse aggregate, sand, Type III cement, water, chemical accelerator

133. Fatigue in steel can be the result of:

(A) a reduction in strength due to cyclical loads

(B) deformation under impact loads

(C) deflection due to overload

(D) expansion due to corrosion

CIVIL AM PRACTICE EXAM

134. Sample concrete cylinders that are 6 inches in diameter and 12 inches high are tested to determine the compressive strength of the concrete f'_c. The test results are as follows:

Sample	Axial Compressive Failure Load (lb)
1	65,447
2	63,617
3	79,168

Based on the above results, the average 28-day compressive strength (psi) is most nearly:

(A) 615
(B) 2,250
(C) 2,450
(D) 2,800

135. During testing of a sample in the laboratory, the following soil data were collected:

Combined weight of compacted soil sample and the mold is 9.11 lb.

Water content of soil sample is 11.5%.

The weight and volume of mold are 4.41 lb and 0.03 ft³, respectively.

The dry unit weight of the soil sample (pcf) is most nearly:

(A) 160
(B) 140
(C) 127
(D) 125

136. Refer to the figure. The net excess excavated material (yd^3) from Station 1+00 to Station 3+00 is most nearly:

(A) 160
(B) 262
(C) 390
(D) 463

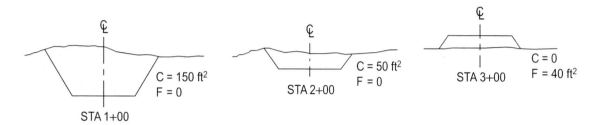

137. An existing pipe connects two maintenance holes (MH). A third MH is planned between the two. At the new MH, the elevation (ft) of the top of the pipe is most nearly:

(A) 623.06
(B) 627.56
(C) 628.06
(D) 628.56

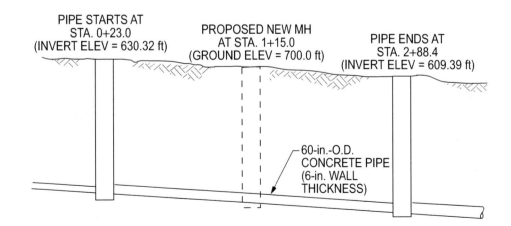

138. Which of the following is **not** a stormwater erosion classification?

(A) Sheet erosion

(B) Rill erosion

(C) Gully erosion

(D) Rushing erosion

139. Based on the soil classification system found in the federal OSHA regulation Subpart P, Excavations, the soil adjacent to an existing building has been classified as Type B. An undisturbed perimeter strip that is 5 ft wide is to be maintained along the face of the building. The excavation is to be 12 ft deep. To meet OSHA excavation requirements, the minimum horizontal distance X (ft) from the toe of the slope to the face of the structure is most nearly:

(A) 11
(B) 14
(C) 17
(D) 23

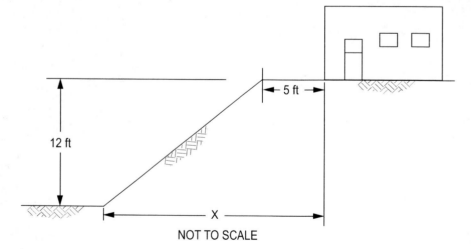

NOT TO SCALE

CIVIL AM PRACTICE EXAM

140. Based on the criteria provided, the steepest backslope (H:V) preferred in the ditch shown is most nearly:

(A) 2:1

(B) 3:1

(C) 5:1

(D) 6:1

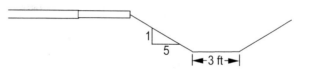

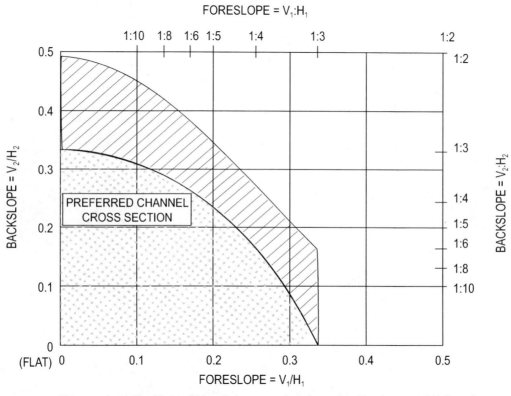

This area is applicable to all Vee ditches, rounded channels with a bottom width less than 2.4 m [8 ft], and trapezoidal channels with bottom widths less than 1.2 m [4 ft].

This area is applicable to rounded channels with bottom width of 2.4 m [8 ft] or more and to trapezoidal channels with bottom widths equal to or greater than 1.2 m [4 ft].

Adapted from AASHTO *Roadside Design Guide*, 4th edition, 2011.

This completes the morning session. Solutions begin on page 63.

GEOTECHNICAL PM PRACTICE EXAM

GEOTECHNICAL PM PRACTICE EXAM

501. The load-settlement behavior of an in situ soil deposit will be determined from laboratory oedometer testing. The type of soil specimen that will provide the most reliable load-deformation characteristics is a soil sample obtained from:

(A) a thin-walled pushed sampler (Shelby tube)

(B) a Standard Penetration Test sampler (SPT or split-spoon sampler)

(C) trimmings from a hand auger

(D) trimmings extracted from the outside of an auger

502. Borings performed on a site indicate the subsurface profile consists of about 30 ft of silty gravel underlain by an interbedded claystone/sandstone bedrock. Which of the following exploration techniques is the best choice to provide an estimated shear wave velocity for the soil and rock profile?

(A) Pressure meter

(B) Seismic refraction

(C) Ground penetrating radar

(D) Electrical resistivity

503. During a Standard Penetration Test (SPT), unusually low blow counts are encountered in a soil expected to be medium-dense to dense sand. This is an indication that the following condition is most likely present:

(A) The sampler drive shoe is badly damaged or worn due to too many drivings to refusal.

(B) Cobbles are encountered.

(C) The sampler drive shoe is plugged.

(D) The groundwater in the borehole is much lower than in situ conditions immediately outside the bore hole.

504. **Figure 1** shows a gradation curve for a soil. Using **Figure 2**, you determine the best USDA textural classification of this soil is:

(A) loam

(B) loamy sand

(C) silt

(D) silty loam

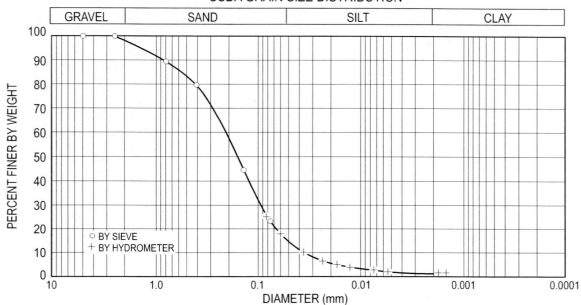

FIGURE 1

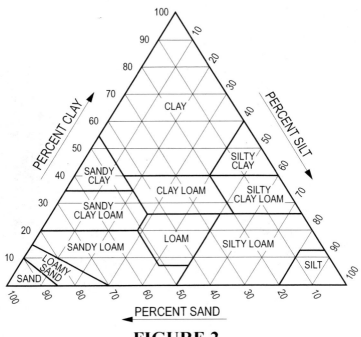

FIGURE 2

GEOTECHNICAL PM PRACTICE EXAM

505. The figure shows a rock core recovered from a depth of 160 ft to 175 ft. Which of the following statements is most correct?

(A) Rock quality designation (RQD) is 94%.

(B) Rock recovery is 80%.

(C) Rock bearing capacity is very poor.

(D) Rock quality designation (RQD) is 80%.

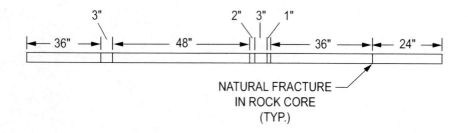

506. The following laboratory test data were obtained for a soil sample:

Sample volume	15.5 cm^3
Sample weight	30.5 g
Dry sample weight	25.0 g
Specific gravity	2.7

The void ratio and saturation of this sample are most nearly:

	Void Ratio	**Saturation**
(A)	0.4	90%
(B)	0.7	100%
(C)	0.4	100%
(D)	0.7	90%

GEOTECHNICAL PM PRACTICE EXAM

507. A saturated cohesionless soil was tested to failure in a triaxial apparatus and the following data recorded:

Maximum vertical test load	121 lb
Sample diameter	3.0 in.
Chamber pressure	16.4 psi
Pore water pressure	10.0 psi

Under drained conditions, the effective friction angle is most nearly:

(A) 25°
(B) 30°
(C) 35°
(D) 40°

508. A consolidated undrained triaxial test is performed on a clay sample. The stress-strain and pore pressure relations are shown. The soil's stress history can be best described as:

(A) overconsolidated
(B) normally consolidated
(C) unconsolidated
(D) remolded

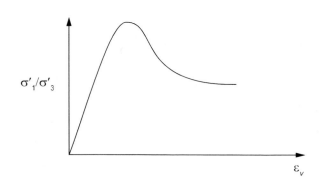

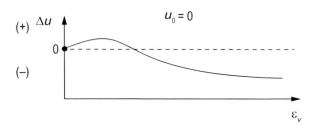

GEOTECHNICAL PM PRACTICE EXAM

509. The following data was collected in a constant-head permeability test:

$L = 20$ in.
$A = 5$ in^2
$h = 30$ in.
$Q = 28$ in^3 over a 3-min period

The hydraulic conductivity (in./sec) of the tested sample is most nearly:

(A) 0.021
(B) 0.62
(C) 1.24
(D) 3.73

GEOTECHNICAL PM PRACTICE EXAM

510. Based on the boring log shown below, the buoyant (submerged) unit weight (pcf) of soil at a depth of 12 ft is most nearly:

(A) 33
(B) 56
(C) 95
(D) 119

DEPTH (ft)	N VALUE	UNIFIED SOIL CLASSIFICATION	DEPTH TO WATER (ft) 5.0	
			MOISTURE CONTENT (%)	DRY DENSITY (pcf)
0	10	SM	10%	105
5 ▽				
6	7	CL	27%	90
10				
12	8	CL	25%	95
15				
19	9	CL	21%	100
20				

GEOTECHNICAL PM PRACTICE EXAM

511. A field density test was performed on compacted fill in accordance with ASTM D 1556 *Standard Test Method for Density and Unit Weight of Soil in Place by the Sand-Cone Method*. The following is a summary of the test data:

Weight of wet soil extracted from the sand-cone hole	6.438 lb
Weight of sand needed to fill the hole (funnel correction already applied)	4.125 lb
The bulk density of the sand used in the sand-cone apparatus	82.4 pcf

Water Content Test:
    Empty cup   0.462 lb
    Cup plus wet soil   1.832 lb
    Cup plus dry soil   1.720 lb

The laboratory maximum dry density performed on the same soil is equal to 130 pcf. Based on this data, the relative compaction of the fill is most nearly:

(A) 99%
(B) 95%
(C) 91%
(D) 63%

GEOTECHNICAL PM PRACTICE EXAM

512. The excavation shown in the figure is required to repair a broken fiber optic line. The excavation is to be open for at least 18 hours. A train will pass the excavation area each hour. As the competent person on-site, you must determine which of the following statements is most correct:

(A) OSHA Soil Type A—Excavation is safe for entry.

(B) OSHA Soil Type A—Excavation should be sloped at 3/4:1 or flatter prior to entry.

(C) OSHA Soil Type B over A— Excavation should be sloped at 3/4:1 in upper 3 ft and 3/4:1 in lower 9 ft prior to entry.

(D) OSHA Soil Type B— Excavation should be sloped at 1:1 or flatter prior to entry.

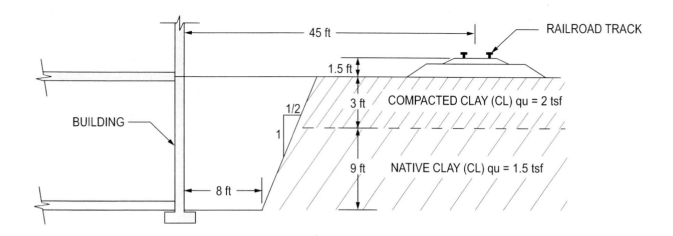

513. A piezometer is to be installed as part of the construction of an earth embankment dam. Which of the following piezometer types would be **least** suitable for measuring excess pore water pressure in a clay soil?

(A) Open-top standpipe

(B) Pneumatic

(C) Vibrating wire

(D) Fiber optic

GEOTECHNICAL PM PRACTICE EXAM

514. The liquefaction potential of a site is to be evaluated. For Layer 3, the design earthquake-induced average shear stress is 450 psf, and the maximum allowable cyclic stress ratio is 0.29. The factor of safety against liquefaction in Layer 3 is most nearly:

(A)   0.7
(B)   1.2
(C)   1.3
(D)   1.4

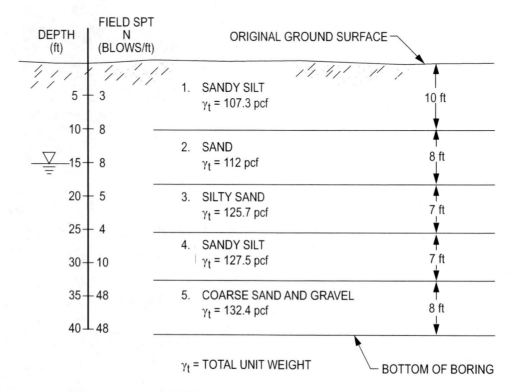

SUBSURFACE INVESTIGATION SUMMARY
NOT TO SCALE

GEOTECHNICAL PM PRACTICE EXAM

515. A building is to be constructed in a seismic zone. A seismic sounding conducted at the site indicated the following shear wave data at various depths.

Depth (ft)	Soil Type	Shear Wave Velocity (ft/sec)
0–10	Fill	500
10–25	Limestone	5,000
25–100	Shale	2,900

What is the site class based on the above profile?

(A) Site Class A

(B) Site Class B

(C) Site Class C

(D) Site Class D

Site Class	Soil Profile Name	Average Properties in Top 100 ft		
		Soil Shear Wave Velocity, $\bar{v}_s$ (ft/s)	Standard Penetration Resistance, $\bar{N}$	Soil Undrained Shear Strength, $\bar{s}_u$ (psf)
A	Hard rock	$\bar{v}_s > 5,000$	Not applicable	Not applicable
B	Rock	$2,500 < \bar{v}_s \leq 5,000$	Not applicable	Not applicable
C	Very dense soil	$1,200 < \bar{v}_s \leq 2,500$	$\bar{N} > 50$	$\bar{s}_u \geq 2,000$
D	Stiff soil profile	$600 \leq \bar{v}_s \leq 1,200$	$15 \leq \bar{N} \leq 50$	$1,000 \leq \bar{s}_u \leq 2,000$
E	Soft soil profile	$\bar{v}_s < 600$	$\bar{N} < 15$	$\bar{s}_u < 1,000$
E		Any profile with more than 10 ft of soil having the following characteristics: 1. Plasticity index $PI > 20$; 2. Moisture content $w \geq 40\%$, and 3. Undrained shear strength $\bar{s}_u < 500$ psf		
F		Any profile containing soils having one or more of the following characteristics: 1. Soils vulnerable to potential failure or collapse under seismic loading such as liquefiable soils, quick and highly sensitive clays, collapsible weakly cemented soils. 2. Peats and/or highly organic clays (H > 10 ft of peat and/or highly organic clay where H = thickness of soil) 3. Very high plasticity clays (H > 25 ft with plasticity index PI > 75) 4. Very thick soft/medium stiff clays (H > 120 ft)		

For SI: 1 foot = 304.8 mm, 1 square foot = 0.0929 m², 1 pound per square foot = 0.0479 kPa.

2000 International Building Code, ©2000, Washington, DC, International Code Council. All rights reserved. www.iccsafe.org.

516. The load test on an infinitely stiff 1-ft × 1-ft plate yielded the applied load vs. deflection diagram shown. The modulus of subgrade reaction (lb/in^3) for a 2-in. deflection of the plate is most nearly:

(A) 8
(B) 10
(C) 70
(D) 105

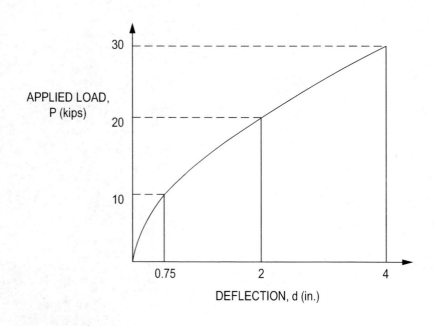

517. A temporary slope will be excavated to the dimensions shown in the figure. Laboratory testing has yielded the geotechnical parameters shown in the chart. The safety factor for the failure surface shown is most nearly:

(A) 1.4
(B) 1.6
(C) 1.8
(D) 2.0

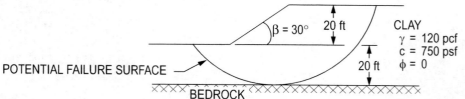

517. (Continued)

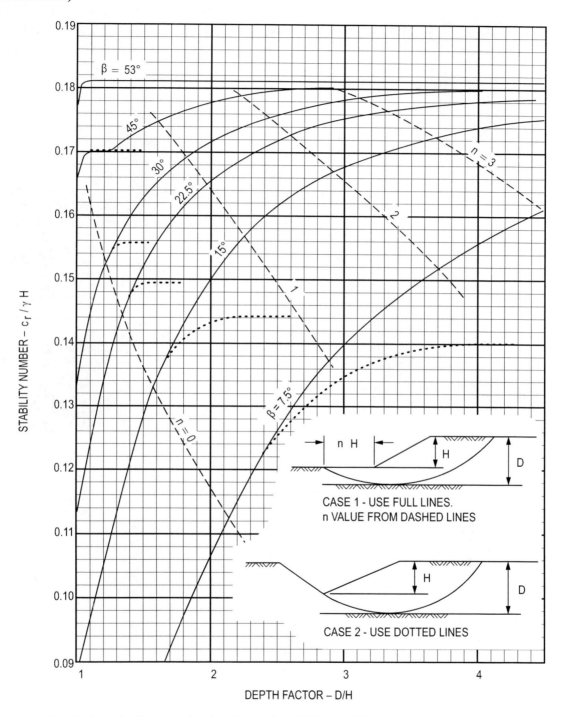

From *Fundamentals of Geotechnical Analysis*, Dunn, Anderson, Riefer, John Wiley and Sons, New York, 1980. Used by permission.

GEOTECHNICAL PM PRACTICE EXAM

518. An earth dam will be built over a pervious layer. The longitudinal length of the pervious layer below the dam is 750 ft, and it is 8 ft thick with a coefficient of permeability of 0.06 in./min. The level of the water at the downstream is maintained at 2,998 ft. The flow of water (gpd) in the pervious material is most nearly:

(A) 1,900
(B) 6,300
(C) 9,500
(D) 14,200

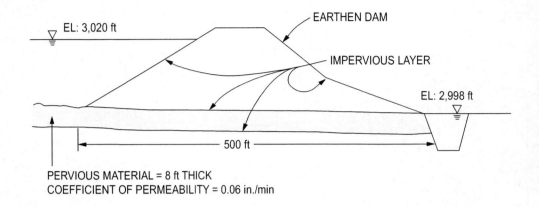

519. An existing asphalt road in very good condition is to be overlain due to increased traffic. The required structural number for the road with overlay is 3.4. Layer coefficients of the materials are as follows:

4 in. of existing asphalt concrete	0.4
8 in. of existing aggregate base course	0.1
Overlay asphaltic concrete	0.42

The required overlay thickness (in.) is most nearly:

(A) 1.5
(B) 2.5
(C) 4.5
(D) 8

520. The figure shows an earth dam where the depth of water behind the dam is 60 ft. Underneath the dam is a 50-ft-thick uniform pervious soil layer that is underlain by impervious soil. Based on the flow net shown, the pore water pressure (psf) at Point A is most nearly:

(A) 7,000
(B) 5,000
(C) 4,000
(D) 3,000

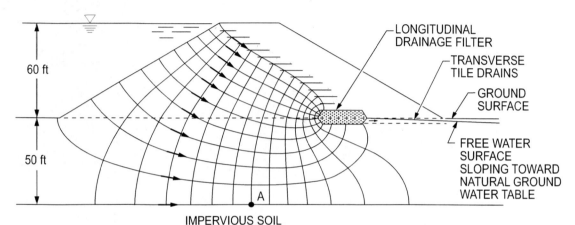

FLOW NET DEPICTING THE FLOW OF WATER THROUGH AN EARTH DAM. (ADAPTED FROM CASAGRANDE, 1940.)

Robert W. Day, *Geotechnical and Foundation Engineering*, McGraw-Hill, ©1999, p. 6.57, Figure 6.26. Used by permission.

521. The figure shows a soil profile where dewatering is taking place over a large area of a site. Ignore the change in the unit weight of the overlying clay. The compression of the sandy silt (in.) in the center of the site due to lowering the piezometric level as shown is most nearly:

(A) 0.03
(B) 0.15
(C) 0.28
(D) 2.8

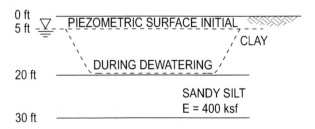

GEOTECHNICAL PM PRACTICE EXAM

522. The figure shows a cross section of a subdrain system that will be used to control the groundwater level beneath a proposed 50-ft-high municipal landfill cell. What type of pipe would be best for the subdrain system?

(A) Sch. 40 PVC pipe that is positioned so that the pipe perforations are located on the bottom of the pipe

(B) Sch. 40 PVC pipe that is constructed as a continuous "tight line" system (i.e., no perforations)

(C) Thin-walled plastic corrugated pipe commonly known as "yard drain" pipe

(D) Sch. 40 PVC pipe that is positioned so that the pipe perforations are located on the top of the pipe

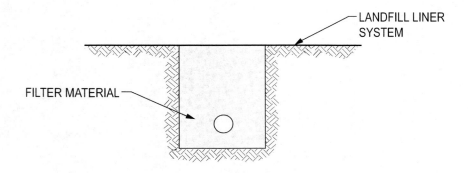

523. For shallow, horizontally stratified, collapsible alluvial soils, which mitigation method to reduce future settlements would be most problematic in an area that is already extensively developed, with nearby existing slab-on-grade structures?

(A) Chemical grouting

(B) Removal of collapsible soil layer

(C) Transfer of load directly to deeper, more stable soils

(D) Prewetting by injecting water into collapsible soil layer

GEOTECHNICAL PM PRACTICE EXAM

524. A laboratory testing program is being developed for a geotechnical investigation in an area suspected of having expansive clay soils. Which of the following laboratory tests would be **least** useful in characterizing the soils as being potentially expansive?

(A) Swell-consolidation
(B) Atterberg limits
(C) Soil suction
(D) Direct shear

525. Structural fill is required to support a slab-on-grade in an area with a deep seasonal frost line. To avoid potential problems with frost heave, the best material for structural fill would be:

(A) low-plasticity cohesive soil compacted dry of optimum (CL)
(B) inelastic silt (ML)
(C) silty sand (SM)
(D) well-graded sand (SW)

526. The Coulomb active earth pressure resultant (lb/ft) for the wall shown in the figure, including the effects of wall friction, is most nearly:

(A) 1,140
(B) 1,279
(C) 1,425
(D) 11,520

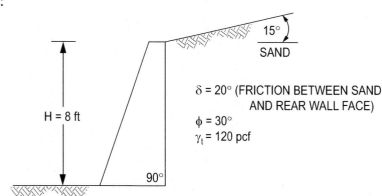

527. The figure shows a cantilever retaining wall subject to a traffic load. Neglect friction between the soil and wall. The total horizontal live load (lb) on the wall per lineal foot is most nearly:

(A) 625
(B) 1,375
(C) 3,125
(D) 6,250

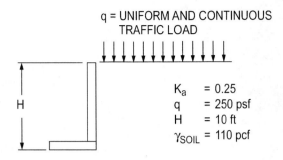

528. The figure shows the placement of an ore stockpile behind an existing cantilevered retaining wall. A structural analysis indicates that no additional lateral earth pressure may be applied to any part of the retaining wall. Using the principle of the active wedge, you determine the minimum distance X (ft) that the ore stockpile must be placed away from the face of the wall is most nearly:

(A) 9
(B) 14
(C) 29
(D) 34

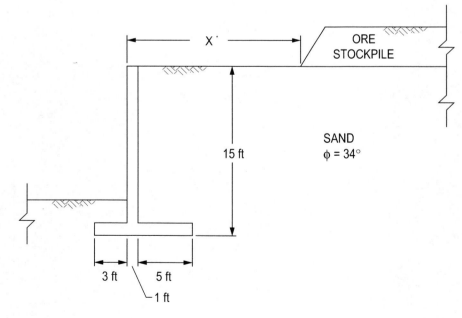

GEOTECHNICAL PM PRACTICE EXAM

529. The figure shows a mechanically stabilized earth retaining wall.

Assumptions:
Friction angle of the backfill, ϕ	30°
Backfill, γ	120 pcf
Coefficient of interaction, C_i	0.7
Embedment length of geogrid layer beyond wall facing, L	8 ft

The ultimate pullout resistance (kips/ft of wall) of the geogrid layer No. 4 is most nearly:

(A) 2.2
(B) 4.4
(C) 6.2
(D) 8.9

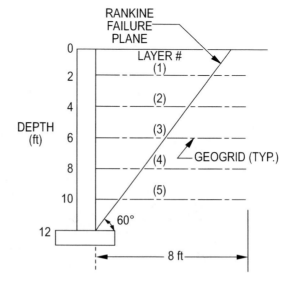

530. As shown in the figure, a horizontal tieback rod and anchor will be used to reduce the bending moments in a retaining wall. The tieback rod has a diameter of 1 in. and a modulus of elasticity of 29,000 ksi. The horizontal force exerted on the tieback rod is 15.5 kips. Neglect friction between the soil and the tieback rod, and assume that the anchor is immobile. If the wall must move laterally by 0.20 in. to reach the active pressure state, the minimum length (ft) of the tieback rod is most nearly:

(A) 15
(B) 25
(C) 32
(D) 300

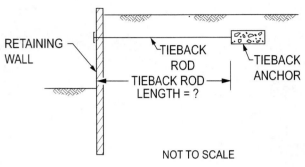

531. Figure 1 shows a foundation. Figure 2 shows bearing capacity factors. The equation for bearing capacity is as follows:

$$q_{ult} = cN_c + \gamma D N_q + 0.5 \gamma B N_\gamma$$

The ultimate bearing capacity (ksf) of the footing is most nearly:

(A) 6
(B) 15
(C) 21
(D) 28

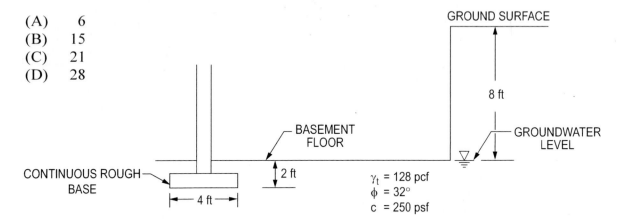

FIGURE 1

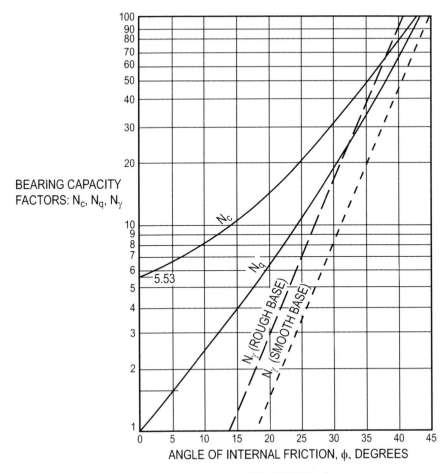

FIGURE 2

532. The effective footing area (ft²) of the eccentrically loaded footing shown in the figure is most nearly:

(A) 4
(B) 7
(C) 14
(D) 24

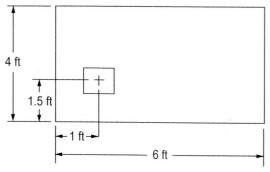

533. A 30-ft × 30-ft square mat foundation will be constructed at ground surface. The subsoil profile is shown in the figure. The mat will apply a uniform pressure of 500 psf. Refer to the chart for stress estimation on the following page. The primary consolidation settlement (in.) of the clay layer directly below the center of the mat is most nearly:

(A) 0.2
(B) 1.0
(C) 2.1
(D) 3.6

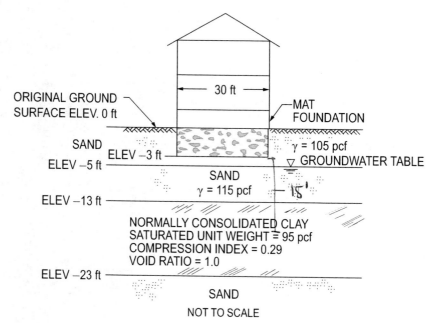

533. (Continued)

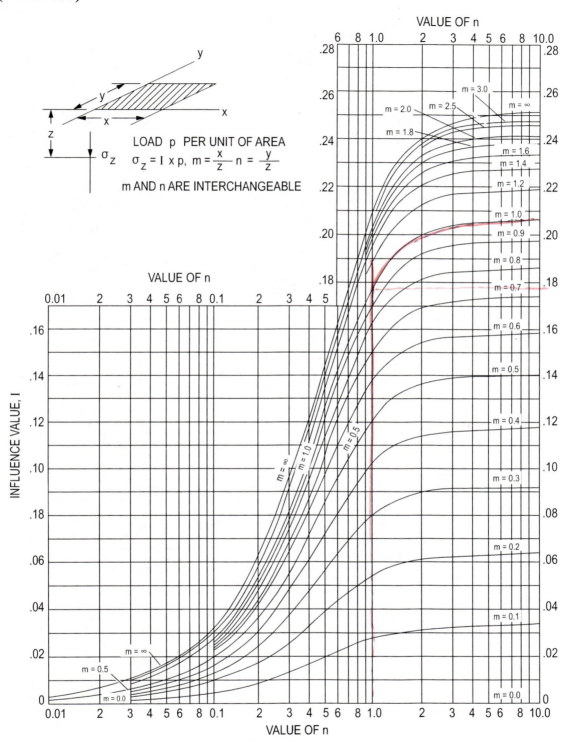

Influence Value for Vertical Stress Beneath a Corner of a
Uniformly Loaded Rectangular Area (Boussinesq Case)

From *Soil Mechanics Design Manual* 7.1, Department of the Navy, Naval Facilities Engineering Command, Virginia, 1982

GEOTECHNICAL PM PRACTICE EXAM

534. A mat foundation will be placed in an excavation to support a building as shown below. The mat foundation dimensions are 100 ft × 200 ft. The building loads are as follows:

Dead load of mat foundation   400 psf
Dead load of building walls, columns, floors, roof, etc.   40,000 kips
Permanent live load   1,500 kips

The net pressure (psf) at the bearing elevation of the mat is most nearly:

(A)   875
(B)   1,100
(C)   1,275
(D)   2,475

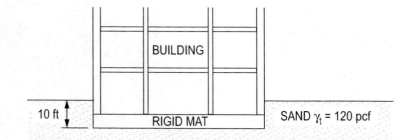

535. The net allowable bearing capacity for the rigid rectangular combined footing shown is 3.0 kips/ft². For the loads and the column spacing shown, size the footing for uniform distribution of soil pressure. The required footing dimensions (B × L) are most nearly:

(A) 5.6 ft × 25.4 ft
(B) 6.1 ft × 23.3 ft
(C) 6.7 ft × 21.0 ft
(D) 7.3 ft × 19.4 ft

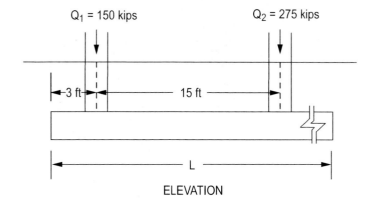

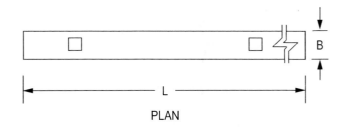

536. A 2-ft × 2-ft square concrete precast pile is shown. The concrete unit weight is 150 pcf. Ignore the resistance of the soft clay but include the weight of the pile. The interface friction angle, δ, on the pile is 0.75 ϕ. For a safety factor of 3.0 against side friction, the allowable tensile capacity (kips) of the pile is most nearly:

(A) 60
(B) 100
(C) 150
(D) 200

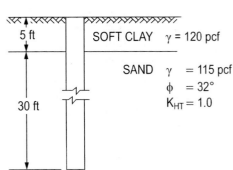

537. A driven steel pile supports an elevated, cast-in-place concrete cap. A horizontal load is applied at the top of the pile. Which option will result in the **least** lateral deflection at the pile cap due to the horizontal load?

(A)
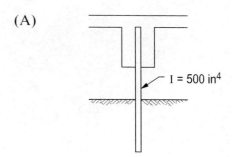
$I = 500$ in^4

(B)
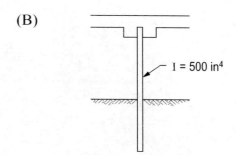
$I = 500$ in^4

(C)
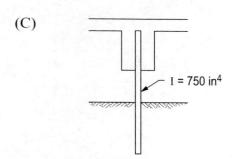
$I = 750$ in^4

(D)
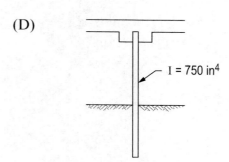
$I = 750$ in^4

538. A pile foundation is shown in the figure. The ultimate side resistance Q_{su}, ultimate point resistance Q_{pu}, and ultimate negative skin friction Q_{nu} are given on the figure. A safety factor of 2.0 is required on the capacity. The safety factor is not applied to downdrag forces. Assume that the downdrag forces reduce the allowable capacity of the pile. The allowable load P (kips) that can be applied to the top of the pile is most nearly:

(A) 115
(B) 135
(C) 155
(D) 310

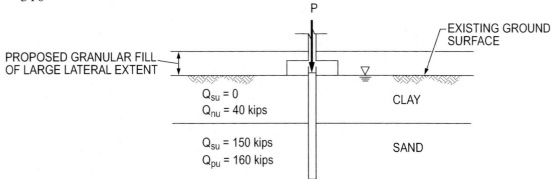

GEOTECHNICAL PM PRACTICE EXAM

539. The figure shows the results of a pile load test performed in accordance with ASTM D1143 *Standard Test Method for Piles Under Static Axial Compressive Load*. A seating load of 18 kips was used for the pile load test. The pile load test was performed on a concrete prestressed pile having a square cross section 1.0 ft × 1.0 ft, length of 40 ft, and modulus of elasticity of 6,000 ksi. Based on the pile load test, the majority of the vertical pile head displacement when loaded from 18 kips to 200 kips is due to:

(A) elastic compression of the pile

(B) plastic compression of the pile

(C) deformation of the pile tip into the bearing strata

(D) deformation of the pile due to shearing of soil along its perimeter

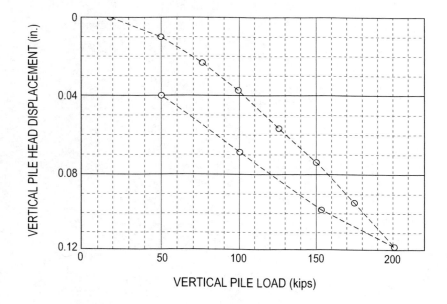

GEOTECHNICAL PM PRACTICE EXAM

540. Load resistance factor design (LRFD) is based on the reliability of the geotechnical data, and accounts for a degree of variability of the measured soil property. What statistical parameters are used to determine the resistance factors for the LRFD method?

(A) Median, standard deviation, coefficient of variation

(B) Median, standard deviation, probability of failure

(C) Mean, standard deviation, coefficient of variation

(D) Mean, coefficient of variation, probability of failure

This completes the afternoon session. Solutions begin on page 83.

CIVIL AM SOLUTIONS

Answers to the Civil AM Practice Exam

Detailed solutions for each question begin on the next page.

101	D	121	A
102	C	122	A
103	C	123	C
104	D	124	B
105	C	125	C
106	B	126	B
107	C	127	C
108	D	128	B
109	B	129	D
110	B	130	D
111	A	131	C
112	A	132	D
113	C	133	A
114	D	134	C
115	B	135	B
116	D	136	C
117	C	137	B
118	D	138	D
119	A	139	C
120	C	140	C

CIVIL AM SOLUTIONS

101. Reference: Peurifoy and Oberlender, *Estimating Construction Costs,* 8th ed., Chapter 10, p. 273, Quantity Takeoff.

$$\text{Horizontal length of side slope} = 14 \times \frac{3}{2} = 21.0 \text{ ft}$$

$$\text{Slope length} = \sqrt{(14)^2 + (21)^2} = 25.24 \text{ ft}$$

$$\text{Cross-sectional area of lining} = [(2 \times 25.24) + 9]\frac{7}{12} = 34.70 \text{ ft}^2$$

$$\text{Volume of lining} = \frac{(34.70 \times 227)}{27} = 291.7 \text{ yd}^3$$

$$\text{Delivered volume} = 291.7 \text{ yd}^3 \times \underset{\text{(waste)}}{1.12} = 327 \text{ yd}^3$$

THE CORRECT ANSWER IS: (D)

102. Reference: Nunnally, *Construction Methods and Management*, 8th ed., 2011, p. 299.

$$D = \frac{\$75,000 - \$10,000}{10}$$

$$D = \$6,500$$

Book value after 8 years = $\$75,000 - (8)(\$6,500) = \$23,000$

THE CORRECT ANSWER IS: (C)

103. Reference: AGC, *Construction Planning and Scheduling*, pub. 3500.1, 6th ed., p. 37.

Crew cost = 2($50/hr) = $100/hr

$$\text{Days allowed} = \frac{\$4,000}{(8 \text{ hr/day})(\$100/\text{hr})} = 5 \text{ days}$$

THE CORRECT ANSWER IS: (C)

CIVIL AM SOLUTIONS

104. Reference: Nunnally, *Construction Methods and Management*, 8th ed., 2011, pp. 282–285.

Activities: ⑦ + ④ + ⑤
Days: 30 + 10 + 10 = 50 days

THE CORRECT ANSWER IS: (D)

105. Reference: Ricketts, Loftin, and Merritt, *Standard Handbook for Civil Engineers*, 5th ed., p. 4.11.

$$1{,}000 \text{ kN} = 1{,}000 \text{ kN} \times \frac{1 \text{ ton}}{8.896444 \text{ kN}} = 112.4 \text{ tons}$$

150 tons > 112.4 tons

THE CORRECT ANSWER IS: (C)

106. Reference: Shapiro, Shapiro, and Shapiro, *Cranes and Derricks*, 3rd ed., 2000, p. 244.

$\tan(x) = \frac{40}{30}$ $x = 53.13°$

$\cos(51.13°) * 100 \text{ ft} = 60 \text{ ft}$

$60 \text{ ft} - 35 \text{ ft} = 25 \text{ ft}$

THE CORRECT ANSWER IS: (B)

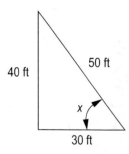

66

CIVIL AM SOLUTIONS

107. Reference: Hurd, *Formwork for Concrete*, ACI SP-4, 7th ed., 2005.

$w = (20 \text{ lb/ft}^2)(8 \text{ ft}) = 160 \text{ lb/vertical ft per brace location}$

$\sum M_a = 0$

$\sum M_a = (160 \text{ lb/ft})(16 \text{ ft})(16 \text{ ft}/2) - 10 \text{ ft } (R_x) = 0$

$R_x = 2,048 \text{ lb}$

Axial load in brace $= \dfrac{(2,048)\sqrt{2}}{1} = 2,896 \text{ lb}$

THE CORRECT ANSWER IS: (C)

108. Reference: NAVFAC, DM 7.2-60.

The wall translation (or strain) required to achieve the passive state is at least twice that required to reach the active state.

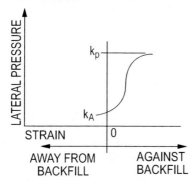

THE CORRECT ANSWER IS: (D)

109. The solution is based on the knowledge that consolidation settlement is the result of the expulsion of pore water from saturated soil due to imposed load. Therefore, the volume of the wick drain effluent (water) to be treated equals the consolidation settlement volume over the affected site area, and is computed as follows:

Affected area	$= 21.5 \text{ acres} \times 43,560 \text{ ft}^2/\text{acre} = 936,540 \text{ ft}^2$
Mean consolidation settlement over affected area	$= 22 \text{ in.} = 1.83 \text{ ft}$
Settlement volume = effluent volume	$= 936,540 \text{ ft}^2 \times 1.83 = 1,713,868 \text{ ft}^3$
Convert to gal: $1,713,868 \text{ ft}^3 \times 7.48 \text{ gal/ft}^3$	$= 12,819,733 \text{ gal}$
Cost for effluent treatment and disposal	$= 12,819,733 \text{ gal} \times \$0.25/\text{gal}$
	$= \$3,204,934$

THE CORRECT ANSWER IS: (B)

CIVIL AM SOLUTIONS

110. Reference: Terzaghi, Peck, Mesri, *Soil Mechanics in Engineering Practice*, 3rd ed., p. 84,

Effective vertical stress at Point A, σ'_v
$= 10\,\text{ft} \times 120\,\text{pcf} + 5\,\text{ft}(120\,\text{pcf} - 62.4\,\text{pcf}) + 7\,\text{ft}(110\,\text{pcf} - 62.4\,\text{pcf})$
$= 1{,}200\,\text{psf} + 288\,\text{psf} + 333\,\text{psf}$
$= 1{,}821\,\text{psf}$

THE CORRECT ANSWER IS: (B)

111. The ultimate bearing capacity would be based on buoyant unit weight, also referred to as the effective unit weight.

Effective unit weight = saturated unit weight − unit weight of water

THE CORRECT ANSWER IS: (A)

112. References: Coduto, *Foundation Design Principles and Practice*, 2nd ed., p. 250.

The long-term settlement for Case I is less than Case II because clay is subject to long-term settlement.

THE CORRECT ANSWER IS: (A)

113. References: Day, *Geotechnical and Foundation Engineering*, 1999, p. 10-27, and NAVFAC 7.1-329.

The minimum factor of safety for permanent slopes is 1.5. Other references use a factor of safety greater than or equal to 1.3, but of the options presented 1.5 is the closest.

THE CORRECT ANSWER IS: (C)

CIVIL AM SOLUTIONS

114. Since the structure is cantilevered, in addition to the wind, dead load and live load will contribute to uplift.

THE CORRECT ANSWER IS: (D)

115. By inspection, Member b = 0 kips, and Member c = 100 kips.

THE CORRECT ANSWER IS: (B)

116. Beam stress, f = M/S, where M = $wL^2/8$ and S = $bh^2/6$.
S is equal for both beams, but M varies because it depends on beam length.
   Beam 1 (shorter beam): $M_1 = wL^2/8$
   Beam 2 (longer beam): $M_2 = w(2L)^2/8 = 4wL^2/8$
M_2 is four times greater than M_1. Therefore the maximum bending stress is four times greater in the longer beam.

THE CORRECT ANSWER IS: (D)

117. Uniform load: $V = \dfrac{wL}{2} = \dfrac{1(30)}{2} = \dfrac{30 \text{ kips}}{2} = 15 \text{ kips}$

Point load: $V = \dfrac{P}{2} = 15 \text{ kips}$
$P = 2(15) = 30 \text{ kips}$

THE CORRECT ANSWER IS: (C)

CIVIL AM SOLUTIONS

118. I_x is maximum for this section by inspection, or calculate $I_x \approx \Sigma Ad^2$ for each section.

THE CORRECT ANSWER IS: (D)

119. $\phi = 32°$ $\quad K_a = \tan^2(45 - \phi/2) = 0.307$

$\gamma_t = 110$ pcf $\quad P_a = (0.5)(110)(8)^2(0.307) = 1,081$ lb/ft

$\quad\quad\quad\quad\quad\quad M_a = (1,081)(8/3) = 2,883$ ft-lb/ft

$(2)(8)(150)(1)(3) = 7,200$ ft-lb/ft $\left.\begin{array}{l}\\ \\\end{array}\right\}$ total $= 8,800$ ft-lb/ft

$(1/2)(2)(8)(150)(1)(2)(2/3) = 1,600$ ft-lb/ft

SF $= 8,800/2,883 = 3.05$

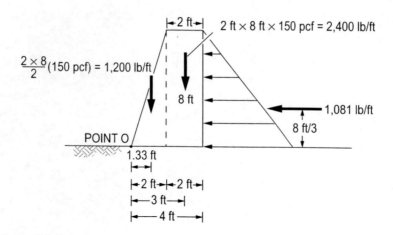

THE CORRECT ANSWER IS: (A)

CIVIL AM SOLUTIONS

120. Reference: Mott, *Applied Fluid Mechanics*, 6th ed., 2005, p. 450.

$$Q = VA = \left\{\frac{1.49}{n}R^{2/3}S^{1/2}\right\}A$$

$$= \left\{\frac{1.49}{0.022}\left[\frac{(1.5\text{ ft} \times 4\text{ ft})}{4\text{ ft} + 2(1.5\text{ ft})}\right]^{2/3}(0.002)^{1/2}\right\}(1.5\text{ ft} \times 4\text{ ft})$$

$$= 16.4 \text{ cfs}$$

Volume $= 25\text{ acre-ft} \times \dfrac{43,560\text{ ft}^3}{1\text{ acre-ft}} = 1.089 \times 10^6\text{ ft}^3$

Time $= \dfrac{1.089 \times 10^6\text{ ft}^3}{16.4\text{ ft}^3/\text{sec}} \times \dfrac{1\text{ min}}{60\text{ sec}} \times \dfrac{1\text{ hr}}{60\text{ min}}$

$= 18.5$ hours

THE CORRECT ANSWER IS: (C)

CIVIL AM SOLUTIONS

121.

Q_1 —— 12 in. 2% ——→ [ ] ——? in. 2%——→ Q_2
Q_1 —— 12 in. 2% ——→

$$2Q_1 = Q_2$$

Reference: Viessman and Lewis, *Introduction to Hydrology*, 4th ed., 1996, p. 252.

$$2[V_1 A_1] = [V_2 A_2]$$

$$2\left[\left(\frac{1.49}{n}\right)(A_1) R_1^{2/3} S^{1/2}\right] = \left[\left(\frac{1.49}{n}\right)(A_2) R_2^{2/3} S^{1/2}\right]$$

$$2\left[(A_1)\left(\frac{A_1}{P_1}\right)^{2/3}\right] = \left[(A_2)\left(\frac{A_2}{P_2}\right)^{2/3}\right]$$

$$A_1 = \frac{\pi D^2}{4} = \frac{\pi (1)^2}{4} = 0.785 \text{ ft}^2$$

$$P_1 = \pi(D) = \pi(1) = 3.14 \text{ ft}$$

$$2\left[(0.785)\left(\frac{0.785}{3.14}\right)^{2/3}\right] = \left[\left(\frac{\pi D_2^2}{4}\right)\left(\frac{\frac{\pi (D_2)^2}{4}}{(\pi D_2)}\right)^{2/3}\right]$$

$$0.623 = \left(\frac{\pi D_2^2}{4}\right)\left(\frac{D_2}{4}\right)^{2/3}$$

$$= \pi\left(\frac{D_2^2}{4}\right)\left(\frac{D_2}{4}\right)^{2/3}$$

$$= \pi(D_2)^{8/3}\left(\frac{1}{4}\right)\left(\frac{1}{4}\right)^{2/3}$$

$$0.623 = 0.311(D_2)^{8/3}$$

$$\left(\frac{0.623}{0.311}\right)^{3/8} = D_2$$

$$D_2 = 1.297 \text{ ft} \times \frac{12 \text{ in.}}{\text{ft}} = 15.6 \text{ in.} \approx 16 \text{ in.}$$

THE CORRECT ANSWER IS: (A)

CIVIL AM SOLUTIONS

122. Reference: Mays, *Water Resources Engineering*, 2001, p. 211.

According to the arithmetic mean method, the average precipitation is simply the average of all the rainfall gages.

Average precipitation = (2.1 + 3.6 + 1.3 + 1.5 + 2.6 + 6.1 + 5.1 + 4.8 + 4.1 + 2.8 + 3.0)/11
Average precipitation = 3.4 in.

THE CORRECT ANSWER IS: (A)

123. Reference: *Water Supply and Pollution Control,* Viessman and Hammer, 6th ed., 1998, p. 229.

From the IDF curve, read a rainfall intensity of 3.5 in./hr for a 50-year frequency rainfall with a 60-min duration.

From the table, the runoff coefficient for a downtown area is 0.70 – 0.95. For the maximum runoff rate, use the high value of 0.95.

$Q = CiA = 0.95 \times 3.5$ in./hr $\times 90$ ac

$Q = 300$ cfs

THE CORRECT ANSWER IS: (C)

124. Reference: Davis and Cornwell, *Introduction to Environmental Engineering*, 4th ed., 2008, p. 61.

$$\text{Time} = \frac{V}{Q}$$

$$V = 400,000 \text{ gal} \times \frac{\text{ft}^3}{7.48 \text{ gal}} = 53,476 \text{ ft}^3$$

$$Q = 1.5 \text{ ft}^3/\text{sec}$$

$$\text{Time} = \frac{53,476 \text{ ft}^3}{1.5 \text{ ft}^3/\text{sec}} \times \frac{1 \text{ hr}}{3,600 \text{ sec}} = 9.9 \text{ hours}$$

THE CORRECT ANSWER IS: (B)

CIVIL AM SOLUTIONS

125. Reference: Merritt, Loftin, and Ricketts, *Standard Handbook for Civil Engineers*, 4th ed., 1996, pp. 21.22 and 21.42.

The Darcy-Weisbach equation is $h_f = f \dfrac{L}{D} \dfrac{V^2}{2g}$

where

h_f = headloss, ft
f = friction factor, unitless
L = length, ft
D = diameter of pipe, ft
V = velocity, ft/sec
g = gravitational constant, 32.2 ft/sec^2

Substituting gives

$$5 \text{ ft} = 0.0115 \times \dfrac{1{,}650 \text{ ft}}{3.0 \text{ ft}} \times \dfrac{V^2}{2 \times 32.2 \text{ ft/sec}^2}$$

$V^2 = 50.91 \text{ ft}^2/\text{sec}^2$

$V = 7.135 \text{ ft/sec}$

$Q = VA = V \times \dfrac{\pi}{4} D^2 = 7.135 \text{ ft/sec} \times \dfrac{\pi}{4}(3.0 \text{ ft})^2$

$Q = 50 \text{ cfs}$

THE CORRECT ANSWER IS: (C)

126. Reference: Lin, Shundar, and C.C. Lee, *Water and Wastewater Calculations Manual*, 2001, p. 240.

$$z_1 + \dfrac{P_1}{\gamma} + \dfrac{v_1^2}{2g} = z_2 + \dfrac{P_2}{\gamma} + \dfrac{v_2^2}{2g}$$

$z_1 = z_2$

Since $A_1 > A_2$, $v_1 < v_2$.

$\therefore \dfrac{v_1^2}{2g} < \dfrac{v_2^2}{2g}$

so $P_1 > P_2$ to balance

THE CORRECT ANSWER IS: (B)

CIVIL AM SOLUTIONS

127. Reference: Hickerson, *Route Location and Design*, 5th ed., p. 64.

$R = 5,729.648/D_C^\circ$
$ = 5,729.648/10 = 572.96 \text{ ft}$

$T = R \tan\left(\dfrac{1}{2}\Delta\right) = R \tan(6.25°)$
$ = 572.96 (\tan 6.25°)$
$ = 572.96 (0.1095178)$
$ = 62.75 \text{ ft}$

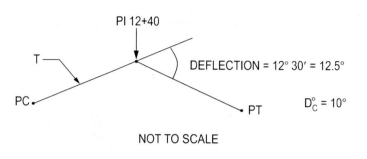

NOT TO SCALE

Station PC = Station PI − T
$ = [12 + 40] - 62.75$
$ = 11 + 77.25$

Station PT = Station PC + length of curve

Length of curve = $L = 100 \Delta/D_C^\circ$
$ = 100(12.5)/10 = 125 \text{ ft}$

Station PT = Station PC + 125 ft = [11 + 77.25] + 125 = 13 + 02.25

THE CORRECT ANSWER IS: (C)

CIVIL AM SOLUTIONS

128. Reference: Hickerson, *Route Location and Design*, 5th ed., pp. 154, 160.

$L = KA$
$K = L/A$
L = length of vertical curve, ft
A = algebraic difference in grades, percent $(g_2 - g_1)$
Given: VPC = 12+00
  VPI = 13+50
  VPT = 15+00
  $g_1 = -2.30\%$
  $g_2 = +3.00\%$
  $L = 300$ ft

$K = \dfrac{L}{A} = \dfrac{300}{3-(-2.3)} = 56.60$ ft/percent for the vertical curve.

The length from Station 14+00 to Station 15+00 = 100 ft

$K = \dfrac{L}{A}$

$A = \dfrac{L}{K} = \dfrac{100}{56.60} = 1.77\%$

$A = g_2 - g_1$

Tangent slope at Station 14+00 = g_1

$g_1 = g_2 - A = 3.00\% - 1.77\% = 1.23\%$

Alternate solution:

Y = elevation at a point X ft from VPC
Y' = slope at a point X ft from VPC
$X = [14+00] - [12+00] = 200$ ft
g_1 = slope 1 in ft/ft
g_2 = slope 2 in ft/ft
L = length of vertical curve, ft

$Y = Y_{VPC} + g_1 X + \left(\dfrac{g_2 - g_1}{2L}\right) X^2$

$Y' = g_1 + \left(\dfrac{g_2 - g_1}{L}\right) X$

$Y' = -0.023 + \left(\dfrac{0.03 - (-0.023)}{300}\right) 200 = 0.0123$ ft/ft or 1.23%

THE CORRECT ANSWER IS: (B)

CIVIL AM SOLUTIONS

129. Reference: Garber and Hoel, *Traffic and Highway Engineering*, 4th ed., pp. 130–132.

$$\text{AADT} = \frac{\sum(\text{Jan. through Dec.})}{12}$$
$$= 833{,}200 / 12 = 69{,}433$$

$\sum(\text{June through Aug.}) = 77{,}300$
$\phantom{\sum(\text{June through Aug.}) =} 78{,}950$
$\phantom{\sum(\text{June through Aug.}) =} \underline{77{,}200}$
$\phantom{\sum(\text{June through Aug.}) =} 233{,}450 / 3 = 77{,}817$

Seasonal factor for June through August
$= 77{,}817 / 69{,}433$
$= 1.121$

THE CORRECT ANSWER IS: (D)

130. Reference: Garber and Hoel, *Traffic & Highway Engineering*, 3rd ed., p. 841.

The commonly used soil classification systems for engineering applications are USCS and AASHTO. Both of these systems use gradation and Atterberg limits as two of the criteria.

THE CORRECT ANSWER IS: (D)

131. Reference: Coduto, Yeung, and Kitch, *Geotechnical Engineering: Principles and Practices*, 2nd ed., p. 184.

The Standard Penetration Test (SPT) N-value provides an indication of the relative density of cohesionless soils.

THE CORRECT ANSWER IS: (C)

CIVIL AM SOLUTIONS

132. Reference: *Design and Control of Concrete Mixtures*, 14th ed., p. 242.

An early-strength concrete is needed with a minimum compressive strength of 3,500 psi. To achieve the requirements, a Type III cement and chemical accelerators would be necessary.

THE CORRECT ANSWER IS: (D)

133. Reference: NCEES, *FE Reference Handbook*, 9.2.

Reduction in strength due to cyclical loads

THE CORRECT ANSWER IS: (A)

134. Area = $\pi d^2/4 = 28$ in^2

Compressive stress = axial load/area

Sample 1 $\quad f'_c = \dfrac{65,447}{28} = 2,313$ psi

Sample 2 $\quad f'_c = \dfrac{63,617}{28} = 2,248$ psi

Sample 3 $\quad f'_c = \dfrac{79,168}{28} = 2,797$ psi

Average $= \dfrac{(2,313 + 2,248 + 2,797)}{3} = 2,452$ psi

THE CORRECT ANSWER IS: (C)

CIVIL AM SOLUTIONS

135. Reference: Garber and Hoel, *Traffic and Highway Engineering,* 4th ed., p. 901.

$$\text{Total density}(\gamma) = \frac{W}{V} = \frac{W_s + W_w}{V_s + V_w + V_a}$$

where γ = total density

W = total weight

V = total volume

W_s = weight soil

W_w = weight of water

V_s = volume of soil

V_w = volume of water

V_a = volume of air

$$\gamma = \frac{9.11 \text{ lb} - 4.41 \text{ lb}}{0.03 \text{ ft}^3} = 156.67 \text{ lb/ft}^3 \text{ (pcf)}$$

Dry unit weight of soil $(\gamma_d) = \frac{\gamma}{1 + w}$

where w = moisture content

$$\gamma_d = \frac{156.67 \text{ pcf}}{1 + 0.115} = 140.51 \text{ pcf}$$

THE CORRECT ANSWER IS: (B)

CIVIL AM SOLUTIONS

136. Reference: Kavanagh, *Surveying with Construction Applications*, 6th ed., 2007, pp. 569–573.

Use Average End Area Method.

Stationing	Excavation (yd^3)	Embankment (yd^3)
1+00 to 2+00	$\dfrac{50+150}{2} \times \dfrac{100}{27} = 370$	
2+00 to 3+00	$\dfrac{50+0}{2} \times \dfrac{100}{27} = 93$	$\dfrac{0+40}{2} \times \dfrac{100}{27} = 74$
Total	463	74

Net excess excavated material = $463 - 74 = 389$ yd^3

THE CORRECT ANSWER IS: (C)

---

137. Reference: Kavanagh, *Surveying with Construction Applications*, 6th ed., 2007, pp. 493–501.

Existing:
$$\Delta H = (2+88.4) - (0+23.0) = 288.4 - 23.0 = 265.4 \text{ ft}$$
$$\Delta V = 630.32 - 609.39 = 20.93 \text{ ft}$$

New:
$$\Delta H = (1+15.0) - (0+23.0) = 115.0 - 23.0 = 92 \text{ ft}$$
$$\Delta V = \frac{92}{265.4} \times 20.93 = 7.26 \text{ ft}$$

Inv Elev. = $630.32 - 7.26 = 623.06$ ft

The top of the pipe will be above the invert elevation by (60 in. − 6 in.)/12 in./ft = 4.50 ft
$623.06 + 4.50 = 627.56$ ft

THE CORRECT ANSWER IS: (B)

CIVIL AM SOLUTIONS

138. Reference: *Developing Your Stormwater Pollution Prevention Plan*, USEPA, May 2007, p. 3. Victor Miguel Ponce, *Engineering Hydrology*, 1st ed., p. 538.

Rushing erosion is not identified in either reference.

THE CORRECT ANSWER IS: (D)

139. Reference: OSHA 29 CFR 1926, Subpart P, Appendix B.

Type B soil has a maximum permissible slope of 1:1.

Therefore, a 12-ft depth requires a 12-ft distance.

Since there is a 5-ft perimeter strip, the minimum distance from the toe of the slope to the face of the structure = 12 ft + 5 ft = 17 ft.

THE CORRECT ANSWER IS: (C)

CIVIL AM SOLUTIONS

140. Reference: AASHTO: *Roadside Design Guide,* 4th ed., 2011, pp. 3-9 and 3-10.

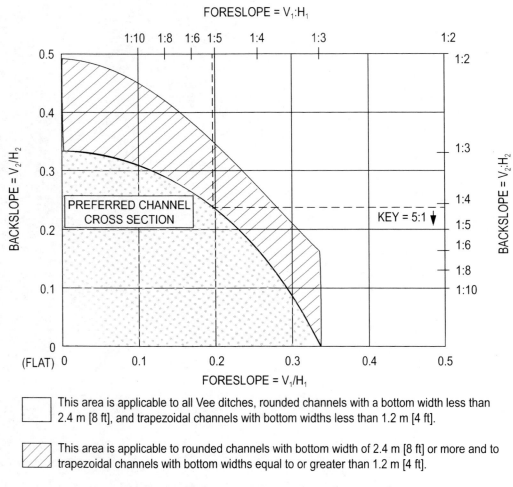

Adapted from AASHTO *Roadside Design Guide,* 4th edition, 2011.

THE CORRECT ANSWER IS: (C)

GEOTECHNICAL PM SOLUTIONS

Answers to the GEOTECHNICAL PM Practice Exam

Detailed solutions for each question begin on the next page.

501	A	521	C
502	B	522	A
503	D	523	D
504	B	524	D
505	D	525	D
506	D	526	C
507	C	527	A
508	A	528	B
509	A	529	B
510	B	530	B
511	C	531	B
512	D	532	B
513	A	533	C
514	C	534	C
515	C	535	A
516	C	536	B
517	C	537	C
518	D	538	A
519	B	539	A
520	B	540	C

GEOTECHNICAL PM SOLUTIONS

501. Reference: Hunt, *Geotechnical Engineering Investigation Manual*, pp. 90–101.

The determination of the load-settlement behavior of an in situ soil deposit is most reliably evaluated by oedometer testing using a soil specimen retrieved with the most minimal disturbance possible. Of the choices given, the least disturbance occurs using a thin-walled pushed sampler.

THE CORRECT ANSWER IS: (A)

502. Reference: McCarthy, *Essentials of Soil Mechanics and Foundations*, 6th ed., pp. 199–211.

Seismic refraction is the best method as it yields the P and S waves for the soil and rock profiles and can be used to evaluate the upper 100 ft of the soil and rock.

THE CORRECT ANSWER IS: (B)

503. Reference: Cheney and Chassie, *Soils and Foundations Workshop Manual*, 2nd ed., 1993, p. 23.

Groundwater coming up into the casing or hollow-stem augers can cause the test zone to become "quick." This loosening will reduce the strength of the soil. The other three conditions typically cause an increase in SPT N-values.

THE CORRECT ANSWER IS: (D)

504. The material percentages are:
  Gravel   2%
  Sand   80%
  Silt   17%
  Clay   1%

Using the USDA chart in Figure 2, loamy sand is the correct answer.

THE CORRECT ANSWER IS: (B)

GEOTECHNICAL PM SOLUTIONS

505. Attempted length = 175 − 160 = 15 ft
Length of core ignoring pieces < 4 in. = 12 ft
RQD = 12/15 = 80%

THE CORRECT ANSWER IS: (D)

506. $V_s = \dfrac{W_s}{G\gamma_w} = \dfrac{25}{2.7 \times 1} = 9.26 \text{ cm}^3$

$V_v = V_t - V_s = 15.5 - 9.26 = 6.24 \text{ cm}^3$

$e = \dfrac{V_v}{V_s} = \dfrac{6.24}{9.26} = 0.67$

$w = \dfrac{W_w}{W_s} \times 100 = \dfrac{30.5 - 25}{25} \times 100 = 22\%$

$Se = Gw$

$S = \dfrac{2.7 \times 0.22}{0.67} \times 100 = 89\%$

Alternatively,

$S = \dfrac{V_w}{V_v} = \dfrac{(30.5 \text{ g} - 25 \text{ g})(1 \text{ cm}^3/\text{g})}{6.24 \text{ cm}^3} \times 100 = 88\%$

THE CORRECT ANSWER IS: (D)

GEOTECHNICAL PM SOLUTIONS

507. Total vertical stress

$$\sigma_1 = \frac{121 \text{ lb}}{\pi(1.5 \text{ in.})^2} + 16.4 = 17.12 \text{ psi} + 16.4 = 33.5 \text{ psi}$$

$$\sigma_3 = 16.4 \text{ psi}, \quad u = 10.0 \text{ psi}$$

$$\sigma'_1 = \sigma_1 - u = 23.5 \text{ psi}, \qquad \sigma'_3 = \sigma_3 - u = 6.4 \text{ psi}$$

$$p = \frac{\sigma'_1 + \sigma'_3}{2} = \frac{23.5 + 6.4}{2} = 14.95 \text{ psi}$$

$$q = \frac{\sigma'_1 - \sigma'_3}{2} = \frac{23.5 - 6.4}{2} = 8.55 \text{ psi}$$

Alternate 1:

$$\alpha' = \arctan\left(\frac{q}{p}\right) = \left(\frac{8.55}{14.95}\right) = 29.8°$$

$$\phi' = \arcsin(\tan \alpha) = (\tan 29.8°) = 34.9°$$

Alternate 2:

$$\phi' = \arcsin\left(\frac{8.55}{14.95}\right) = 34.9°$$

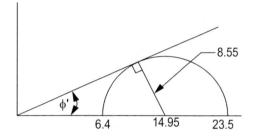

THE CORRECT ANSWER IS: (C)

508. Reference: Holtz, Kovacs, and Sheahan, *Introduction to Geotechnical Engineering*, 2nd ed., p. 572.

Overconsolidated behavior yields a distinct peak in the stress-strain curve and an initial increase in pore pressure followed by reduction due to dilation.

THE CORRECT ANSWER IS: (A)

GEOTECHNICAL PM SOLUTIONS

509. Reference: Das, *Principles of Geotechnical Engineering*, 6th ed., p. 162.

$$K = \frac{QL}{Aht}$$

$$= \frac{28 \text{ in}^3 \times 20 \text{ in.}}{5 \text{ in}^2 \times 30 \text{ in.} \times (3 \text{ min} \times 60 \text{ sec/min})}$$

$$= 0.021 \text{ in./sec}$$

THE CORRECT ANSWER IS: (A)

510. Assume soil below water table is saturated and $\gamma = \gamma_{sat} = \gamma_{dry}(1 + w)$.

$\gamma_{sat} = 95(1 + 0.25) = 118.8$ pcf

$\gamma_b = \gamma_{sat} - \gamma_w$
$ = 118 - 62.4 = 56.4$ pcf

THE CORRECT ANSWER IS: (B)

511. Volume of hole = 4.125 lb/82.4 pcf = 0.0501 cf

$\gamma_t = 6.438$ lb/0.0501 cf = 128.6 pcf

$w = (1.832 - 1.720)/(1.720 - 0.462) = 0.112/1.258 = 0.0890$

$\gamma_d = 128.6/(1 + 0.0890) = 118.0$

Relative compaction = $\gamma_d/\gamma_{d\,max} = 118.0/130.0 = 0.908$ or 91%

THE CORRECT ANSWER IS: (C)

512. Reference: OSHA 29 CFR Part 1926, Subpart P, Appendix A and B.

Per definition for Soil Type A Exception (ii), the soil cannot be classified as "A" as it will be subject to vibration from the railroad. Therefore, per definition of Soil Type B (iv), the soil should be classified as "B." Per Table B-1, Type B soils should be sloped at 1:1 or flatter.

THE CORRECT ANSWER IS: (D)

GEOTECHNICAL PM SOLUTIONS

513. Reference: Fang, *Foundation Engineering Handbook*, 2nd ed., p. 433; and McCarthy, *Essentials of Soil Mechanics and Foundation*, 6th ed., pp. 181–183.

Open-top pipe is most appropriate for coarse soil types with high permeability. It is not suitable for clay because of lag time and is susceptible to shearing of sides of bore hole.

THE CORRECT ANSWER IS: (A)

514. Effective vertical stress at middle of Layer 3
$$\sigma'_v = (107.3 \times 10) + (112 \times 5) + (112 - 62.4) \times 3 + (125.7 - 62.4) \times 3.5 = 2,003 \text{ psf}$$

Shear stress to cause liquefaction
$$\Sigma' = \sigma'_v \times \text{CSR} = 2,003 \times 0.29 = 580.9 \text{ psf}$$

Determine factor of safety against liquefaction in Layer 3
$$\text{FS} = \frac{\Sigma'}{\Sigma} = \frac{580.9}{450} = 1.29$$

THE CORRECT ANSWER IS: (C)

515.
$$\frac{100}{\dfrac{10}{500} + \dfrac{15}{5,000} + \dfrac{75}{2,900}} = \frac{100}{0.02 + 0.003 + 0.026}$$

$$= \frac{100}{0.049} = 2,040$$

$$1,200 < 2,040 < 2,500 \quad \therefore \text{Site Class C}$$

THE CORRECT ANSWER IS: (C)

GEOTECHNICAL PM SOLUTIONS

516. Reference: Bowles, *Foundation Analysis & Design*, 5th ed., p. 501.

According to the diagram, a 2-in. deflection yields an applied load, P = 20,000 lb.

$$q = \frac{P}{A} = \frac{20,000}{144} = 139 \text{ lb/in}^2$$

Modulus of subgrade reaction, $K_g = \frac{q}{d} = \frac{139 \text{ lb/in}^2}{2 \text{ in.}} = 69.5 \text{ lb/in}^3$

THE CORRECT ANSWER IS: (C)

517. $D = 20$
$\left.\begin{array}{l} D/H = 2 \\ \beta = 30 \end{array}\right\}$ stability number $= 0.172 = \dfrac{c_r}{\gamma H} = \dfrac{c_r}{120 \times 20}$

$c_r \cong 413$

$FS = \dfrac{c}{c_r} = \dfrac{750}{413} = 1.82$

THE CORRECT ANSWER IS: (C)

518. Reference: Holtz, Kovacs, and Sheahan, *An Introduction to Geotechnical Engineering*, 2nd ed., p. 275.

$q = k \, i \, a$

where

a = c/s area = $8 \times 750 = 6,000$ sq ft

i = hydraulic gradient

$\quad = \dfrac{3,020 - 2,998}{500} = 0.044$

$k = 0.06$ in./min = $0.06 \times 1/12$ ft/min

$\quad = 0.005$ ft/min

$= 0.005$ ft/min $\times 0.044 \times 6,000$ ft^2

$q = 1.32$ ft^3/min

$\quad = 1.32$ ft$^3 \times 24 \times 60$ ft^3/day

$\quad = 1,900.8$ ft^3/day

$q = 1,900.8 \times 7.48$ gal/day

$q = 14,220$ gal/day

THE CORRECT ANSWER IS: (D)

GEOTECHNICAL PM SOLUTIONS

519. (4-in. asphalt concrete)(0.4) =   1.6
(8-in. aggregate base course)(0.1) =   0.8
Total   2.4

Required overlay $= \dfrac{3.4 - 2.4}{0.42} = 2.4$ in.

THE CORRECT ANSWER IS: (B)

520. Drops at Point A   7
Total number of drops   14
Head loss between up and down grad sides   60 ft

$\Delta h = \left(\dfrac{7}{14}\right)(60) = 30$ ft

At Point A:

$u = (110\text{ ft} - 30\text{ ft})(62.4\text{ pcf})$
$ = (80\text{ ft})(62.4\text{ pcf})$
$ = 4{,}992$ psf

THE CORRECT ANSWER IS: (B)

GEOTECHNICAL PM SOLUTIONS

521. Before:
$$\sigma_A = 5\gamma_{C1} + 15\gamma_{C2} + 5\gamma_S \qquad \gamma_{C1} = \gamma_{C2} = \gamma_C$$
$$\mu_A = (15+5)\gamma_W = 20\gamma_W$$
$$\bar{\sigma}_{Ai} = \sigma_A - \mu_A = 20\gamma_C + 5\gamma_S - 20\gamma_W$$

After:
$$\sigma_A = 20\gamma_C + 5\gamma_S$$
$$\mu_A = 5\gamma_W$$
$$\bar{\sigma}_{Af} = 20\gamma_C + 5\gamma_S - 5\gamma_W$$
$$\Delta\bar{\sigma}_A = 15\gamma_W \qquad \text{This is the change in stress across the entire sandy silt layer.}$$
$$\delta = \frac{\Delta\bar{\sigma}_A}{E} \times H_{SAND} = \frac{(15)(62.4\,\text{psf})}{400{,}000\,\text{psf}} \times (10 \times 12)\,\text{in.} = 0.28\,\text{in.}$$

```
0 ft  ───────────────────────        0 ft  ───────────────────
5 ft  -----▽---- γ_C1
                      CLAY
           γ_C2
20 ft ───────────────────────        20 ft ─────────────▽─────
           γ_S  + A   SANDY SILT              + A
30 ft ───────────────────────        30 ft ───────────────────
              BEFORE                             AFTER
```

THE CORRECT ANSWER IS: (C)

522. Reference: Day, *Geotechnical Engineering Portable Handbook*, 2000, p. B.15.

Note 3 states: "perforations placed on underside of pipe"

Reasons:
1. Allows water to quickly enter pipe.
2. Do not want trench to fill up all the way to the top of pipe.
3. Do not want entire pipe to be submerged in effluent.

THE CORRECT ANSWER IS: (A)

GEOTECHNICAL PM SOLUTIONS

523. Reference: Coduto, *Foundation Design Principles and Practices*, 2nd ed., pp. 715–718.

The horizontally stratified alluvial soil may flow horizontally more than it does in the vertical direction. Due to the nearby development, this method may induce settlement of these structures as the collapsible soil under them may collapse because of the water that was injected.

THE CORRECT ANSWER IS: (D)

524. Reference: Terzaghi, Pack, and Mesri, *Soil Mechanics in Engineering Practice*, 3rd ed., pp. 116–121.

Direct shear would give ϕ and C but would not give anything related to expansive soils.

THE CORRECT ANSWER IS: (D)

525. Well-graded sand is the least susceptible to frost heave.

THE CORRECT ANSWER IS: (D)

526.
$$K_a = \frac{\sin^2(\alpha + \phi)}{\sin^2\alpha \sin(\alpha - \delta)\left[1 + \sqrt{\frac{\sin(\phi + \delta)\sin(\phi - \beta)}{\sin(\alpha - \delta)\sin(\alpha + \beta)}}\right]^2}$$
$$= 0.371$$

Also refer to earth pressure charts.

$$P_a = \frac{\gamma H^2}{2} K_a$$
$$= \frac{(120 \text{ lb/ft}^3)(8 \text{ ft})^2}{2}(0.371)$$
$$= 1,425 \text{ lb/ft}$$

THE CORRECT ANSWER IS: (C)

GEOTECHNICAL PM SOLUTIONS

527. $P_{a\ live} = qHK_a = (250)(10)(0.25) = 625$ lb/ft

THE CORRECT ANSWER IS: (A)

528. Assume active earth pressure acts on a vertical plane through the heel of the wall, with active failure plane inclined as indicated:

$$X = 1\ ft + 5\ ft + \frac{15}{\tan(45 + 34/2)}$$
$$= 14\ ft$$

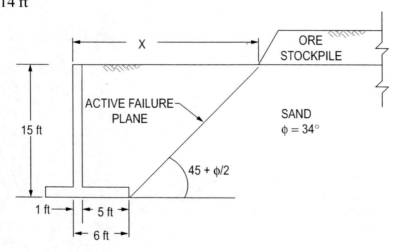

THE CORRECT ANSWER IS: (B)

529. Effective length L of geogrid #4 behind failure wedge = $8\ ft - (4ft)(\tan 30°) = 5.69$ ft.
The vertical stress σ_v acting on the geogrid = $(120\ pcf)(8\ ft) = 960$ psf.
The geogrid obtains resistance from both its top and bottom, and including interaction coefficient C_i:

$$P = 2L\sigma_v C_i \tan\phi$$
$$= (2)(5.69\ ft)(960\ psf)(0.7)(\tan 30°)$$
$$= 4,415\ lb/ft = 4.4\ kips/ft$$

THE CORRECT ANSWER IS: (B)

GEOTECHNICAL PM SOLUTIONS

530. $\Delta = \dfrac{PL}{AE}$

$0.20 \text{ in.} = \dfrac{(15.5 \text{ kips}) L}{\pi (0.5)^2 \, 29{,}000 \text{ ksi}}$ or $L = 294 \text{ in.} = 24.5 \text{ ft}$

THE CORRECT ANSWER IS: (B)

531. $q_{ult} = cN_c + \gamma DN_q + 1/2 \, \gamma BN_\gamma$
$= (250)(36) + \big[(128 - 62.4)(2 \text{ ft})(24)\big] + \big[(0.5)(128 - 62.4)(4)(25)\big]$
$= 9{,}000 + 3{,}148 + 3{,}280$
$= 15{,}428 \text{ psf} = 15.4 \text{ ksf}$

THE CORRECT ANSWER IS: (B)

532. Reference: Das, *Principles of Foundation Engineering*, 6th ed., pp. 150–151.

$e_L = 3 - 1 = 2$ $\quad$ $e_B = 2 - 1.5 = 0.5$
$e_L/L = 2/6 = 0.33$ $\quad$ $e_B/B = 0.5/4 = 0.125$

Case II, Figure 3.16
$L_1/L \approx 0.5$ $\quad$ $L_2/L = 0.1$
$L_1/6 = 0.5$ $\quad$ $L_2/6 = 0.1$
$L_1 = 3.0$ $\quad$ $L_2 = 0.6$
$A' = 1/2(L_1 + L_2)B$
$ = 1/2(3.0 + 0.6)4$
$ = 7.2 \text{ ft}^2$

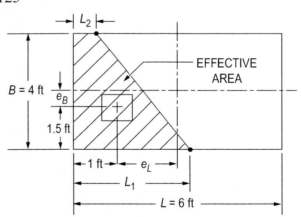

Alternate Solution:

$A = 2(1.5) \times 2(1.0)$
$ = 6.0 \text{ ft}^2$

THE CORRECT ANSWER IS: (B)

GEOTECHNICAL PM SOLUTIONS

533. Determine $\Delta\sigma$ from foundation using the chart provided.

From chart, $x = 15$, $y = 15$, $z = 15$ — center of clay layer
so $m = 1$, $n = 1$ — at corners
$I = 0.18$ Total $I_T = 4(I) = 4(0.18) = 0.72$
$P = 500$ $\Delta\sigma = 0.72(500) = 360$ psf
$\sigma'_{vo} = 5(105) + 8(115 - 62.4) + 5(95 - 62.4) = 1{,}109$ psf

$$S = C_c \frac{H_o}{1 + e_o} \log\left(\frac{\sigma'_{vo} + \Delta\sigma}{\sigma'_{vo}}\right)$$ — half of clay layer

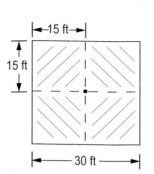

$$S = 0.29 \frac{10(12)}{1 + 1.0} \log\left(\frac{1{,}109 + 360}{1{,}109}\right)$$

$S = 2.1$ in.

Use superposition to determine change in stress in center of 30-ft mat from chart. Determine I below corner and then multiply by four.

THE CORRECT ANSWER IS: (C)

---

534. • Relief due to excavation $= (-10 \text{ ft})(120 \text{ pcf}) = -1{,}200$ psf
Dead load of mat foundation (given) $= 400$ psf

Dead load of building $= \dfrac{40{,}000 \text{ kips}}{(100 \text{ ft})(200 \text{ ft})} \times \dfrac{1{,}000 \text{ lb}}{\text{kip}} = 2{,}000$ psf

Permanent live load $= \dfrac{1{,}500 \text{ kips}}{(100 \text{ ft})(200 \text{ ft})} \times \dfrac{1{,}000 \text{ lb}}{\text{kip}} = 75$ psf

Bring everything to psf

Net stress at bearing elevation $= 400 + 2{,}000 + 75 - 1{,}200 = 1{,}275$ psf

THE CORRECT ANSWER IS: (C)

GEOTECHNICAL PM SOLUTIONS

535. Foundation area, $A = \dfrac{150 \text{ kips} + 275 \text{ kips}}{3.0 \text{ kips/ft}^2} = 141.7 \text{ ft}^2$

Location of column load resultant (centroid) $= \dfrac{275 \text{ kips} \times 15 \text{ ft}}{150 \text{ kips} \times 275 \text{ kips}} = 9.7 \text{ ft right of } Q_1$

For uniform distribution of soil pressure, the loads pass through the foundation centroid.

$L = 2(3 \text{ ft} + 9.7 \text{ ft}) = 25.4 \text{ ft}$

$B = \dfrac{A}{L} = \dfrac{141.7 \text{ ft}^2}{25.4 \text{ ft}} = 5.6 \text{ ft}$

Footing dimensions = 5.6 ft × 25.4 ft

THE CORRECT ANSWER IS: (A)

536. $T_{all} = \dfrac{T_{ult}}{FS} + W_{pile}$

$T_{ult} = K_H \, P_o \, \tan\delta \times H \times S$

Surface area, $S = 2 \text{ ft} \times 4 = 8 \text{ ft/lf}$

Pile embedment, $H = 30 \text{ ft}$

$\tan\delta = \tan(0.75 \times 32°) = 0.445$

Average overburden pressure, $P_o = 2{,}325 \text{ psf}$

at 5 ft $\quad \sigma_u = 5(120) = 600$

at 35 ft $\quad \sigma_u = 600 + 30(115) = 4{,}050$

$T_{ult} = 1.0 \times 2{,}325 \times 0.445 \times 30 \text{ ft} \times 8 \text{ ft} = 248 \text{ kips}$

$W_{pile} = 2 \text{ ft} \times 2 \text{ ft} \times \dfrac{150}{1{,}000} \times 35 = 21 \text{ kips}$

$T_{all} = \dfrac{248}{3} + 21 = 104 \text{ kips}$

THE CORRECT ANSWER IS: (B)

GEOTECHNICAL PM SOLUTIONS

537. A higher I value results in a stiffer pile and less deflection. Greater embedment in the pile cap is closer to fixed conditions and gives less deflection than the pinned condition.

THE CORRECT ANSWER IS: (C)

538. Reference: Das, *Principles of Foundation Engineering*, 4th ed., p 350, eq. 8.8.

$$Q_{ult} = Q_{su} + Q_{pu}$$
$$Q_{ult} = 150 + 160 = 310 \text{ kips}$$
$$Q_{all} = \frac{Q_{ult}}{F.S.} - Q_{nu} \qquad \text{NAFVAC DM 7.2, p. 7.2-211, eq. 5d}$$
$$= \frac{310}{2} - 40$$
$$= 115 \text{ kips}$$

THE CORRECT ANSWER IS: (A)

539. $\Delta = \frac{PL}{AE}$; $P = 200 \text{ kips} - 18 \text{ kips} = 182 \text{ kips}$

$$\Delta = \frac{182 \text{ kips } (40 \text{ ft})}{1 \text{ ft}^2 (6,000 \text{ kips/in}^2)(144 \text{ in}^2/\text{ft}^2)}$$
$$= 8.43 \times 10^{-3} \text{ ft}$$
$$= 0.1 \text{ in.}$$

Since pile head deformation is 0.12 in. from 18 kips to 200 kips, the majority of the vertical pile head displacement (i.e., 0.10/0.12 = 83%) is due to elastic compression of the pile. This is also evident from the unload portion of the curve, where most of the pile head displacement is recovered upon unloading.

THE CORRECT ANSWER IS: (A)

GEOTECHNICAL PM SOLUTIONS

540. Reference: FHWA GEC-10, LRFD *Drilled Shafts; Soils & Foundations Reference Manual*, NHI-10-016, NHI-06-089, p. 3-31; App. C.

Mean (average) value with standard deviation and coefficient of variation.

THE CORRECT ANSWER IS: (C)

PE Practice Exams Published by NCEES

Chemical
Civil: Construction
Civil: Structural
Civil: Transportation
Civil: Water Resources and Environmental
Electrical and Computer: Electrical and Electronics
Electrical and Computer: Computer Engineering
Electrical and Computer: Power
Environmental
Mechanical: HVAC and Refrigeration
Mechanical: Mechanical Systems and Materials
Mechanical: Thermal and Fluids Systems
Structural

For more information about these and other NCEES publications and services, visit NCEES.org or call Client Services at (800) 250-3196.